SpringerBriefs in Computer Science

For further volumes:
http://www.springer.com/series/10028

Peng Yue

Semantic Web-based Intelligent Geospatial Web Services

Peng Yue
State Key Laboratory of Information
Engineering in Surveying, Mapping and
Remote Sensing (LIESMARS)
Wuhan University
Wuhan, Hubei
People's Republic of China

ISSN 2191-5768 ISSN 2191-5776 (electronic)
ISBN 978-1-4614-6808-0 ISBN 978-1-4614-6809-7 (eBook)
DOI 10.1007/978-1-4614-6809-7
Springer New York Heidelberg Dordrecht London

Library of Congress Control Number: 2013935482

Printed on acid-free paper

Springer is part of Springer Science+Business Media (www.springer.com)

Preface

In the twenty-first century, when rapidly evolving information technologies (IT) and increasing Earth observing capabilities affect nearly every aspect of geospatial information production, management, and consumption, we have seen a growing trend in wide sharing and distributed processing of geospatial information using Web Services. Geospatial service, or named geographic information services (GIService), has become an important topic in geographic information systems (GIS).

The latest Earth observing technologies and systems have provided large volumes of geospatial data. We are experiencing a data-rich yet analysis-poor period. Geospatial users need "intelligence" in data discovery, information extraction, and knowledge discovery using GIS including GIService. The book leverages Semantic Web technologies and GIService to provide intelligent geospatial Web services. The contents cover semantics for geospatial data and services, semantics-enhanced geospatial catalogue services, and intelligent geoprocessing service chaining. The technologies and approaches are layered on existing interoperability standards (OGC, W3C, and OASIS) in both geospatial and general information domains.

The work in this book started in 2004, when I conducted my doctoral research at George Mason University in Fairfax County, Virginia, USA. The book summarizes the work in the past several years. We hope that this material will be of interest to many others who are either GIS or IT specialists, and help to strengthen wide applications and services of geospatial information.

Finally, I would like to thank Prof. Liping Di and Jianya Gong for their constructive comments and suggestions to my research. The research activities in the book have been funded by the National Basic Research Program of China (2011CB707105), the National Natural Science Foundation of China (41271397 and 41023001), and LIESMARS Special Research Funding. I am also grateful to Springer and Courtney Clark, the editor from the Computer Science department at Springer, for inviting and helping me to publish the book.

December 2012 Peng Yue

Preface

Contents

Chapter 1
Introduction

1.1 Context

More than 150 Earth observation satellites are currently in orbit measuring the state of the Earth system (Tatem et al. 2008). These satellites, together with countless air-, land-, and water-based monitoring systems, are generating large volumes of geospatial data. For example, the National Aeronautics and Space Administration (NASA)'s Earth Observing System (EOS) alone collect 1000 terabytes annually (Clery and Voss 2005). This unprecedented data-collecting capability brings considerable challenges to geospatial research and applications, one of which is how to derive high-level information and knowledge from the oceans of data in an effective and timely way. The traditional methods of analyzing data by expert analysts fall far short of today's increased demands for geospatial knowledge. As a result, much data may never been analyzed even once after collection. Geospatial users are experiencing a data-rich yet analysis-poor period. Therefore, technologies for semi-automated or automated geospatial knowledge discovery and dissemination are urgently needed for geospatial applications.

A new information infrastructure, the so-called Cyberinfrastructure (in the United States) or e-Infrastructure (in Europe) (Hey and Trefethen 2005), is being developed to support the next generation of geoscientific research. The Cyberinfrastructure will be a comprehensive information infrastructure that integrates computing hardware and systems, data and information resources, networks, digitally enabled-sensors, online instruments and observatories, virtual organizations, and experimental facilities, along with an interoperable suite of software and middleware tools and services (NSF 2007). With this information architecture, large volumes of data and powerful computing resources are available to all users, thus significantly enhancing their ability to use online/near-line data over the Web and allowing the widespread automation of data analysis and computation. Scientists can use services to contribute their original content or value-added products to the community. This cyber community will evolve and become a collective knowledge base.

P. Yue, *Semantic Web-based Intelligent Geospatial Web Services*,
SpringerBriefs in Computer Science, DOI: 10.1007/978-1-4614-6809-7_1,

With the advancement of Cyberinfrastructure, Foster (2005) uses the term *Service-Oriented Science* to refer to the scientific research supported by distributed networks of interoperating services. Service-Oriented Architecture (SOA) has shown prospects for providing valuable geospatial data and processing functions for worldwide open use. SOA is a way of reorganizing a portfolio of previously siloed software applications and supporting infrastructure into an inter-connected set of services, each accessible through standard interfaces and messaging protocols (Papazoglou 2003). In fact, service technologies has been explored across multiple disciplines in different countries, such as the European Commission Ground European Network for Earth Science Interoperations-Digital Earth Communities (GENESI-DEC) project (GENESI-DEC 2012), the Global Earth Observation System of Systems (GEOSS) (GEOSS 2012), the UK e-science program (Hey and Trefethen 2005), the U.S. NASA GES-DISC (Goddard Earth Sciences Data and Information Services Center) Interactive Online Visualization ANd aNalysis Infrastructure (Giovanni) (Berrick et al. 2009), the European Space Agency (ESA) Grid Processing on Demand (G-POD) (ESA 2012), and U.S. National Science Foundation (NSF) funded GEON project (GEON 2003) and EarthCube (EarthCube 2011).

Web Service technologies are a set of technologies for the implementation of SOA. They allow scientists to set up this infrastructure for collaborative sharing of such distributed resources as geospatial data, processing modules, and process models, and are the technologies most widely used to support the Cyberinfrastructure. A Web Service is a software system designed to support interoperable machine-to-machine interaction over a network (Booth et al. 2004). It has a standard interface to enable the interoperation of different software systems, so that Web Services developed by different organizations can be combined to fulfill users' requests. The interoperable services can be published, discovered, chained, and executed through the Web. A number of interoperable services have been available in the geospatial community, most notably the Open Geospatial Consortium (OGC) standard-compliant services, including Web Feature Service (WFS) (Vretanos 2010), Web Map Service (WMS) (de la Beaujardière 2006), Web Coverage Service (WCS) (Baumann 2010), Sensor Observation Service (SOS) (Bröring et al. 2012), Catalogue Services for Web (CSW) (Nebert et al. 2007), and Web Processing Service (WPS) (Schut 2007).

While Web Service technologies are used widely in the geospatial domain, there are substantial research challenges to develop high-level intelligent middleware services and domain-specific services for problem-solving and scientific discovery in the Cyberinfrastructure (Hey and Trefethen 2005). The traditional focus on discovery of and access to geospatial data is being expanded primarily to enable scientific research using the Cyberinfrastructure, with its heavy analysis and synthesis demands (Brodaric et al. 2009). Typical activities involve distributed geoprocessing workflows that support information processing and knowledge discovery from vast, heterogeneous data sets. Users need capabilities on dynamically and collaboratively developing interoperable, Web-executable geospatial service modules and models, and applying them on-line to any part of the peta-byte archives to obtain customized information products rather than only raw data (Di 2004).

1.2 Motivating Examples

The application areas of geospatial data are diverse, such as meteorology, global climate change, agriculture, forestry, flood monitoring, wildfire detection and monitoring, geology, oil spill detection and monitoring. Often, these applications are both computing- and data-intensive and involve diverse sources of data and complex processing functions. The data obtained from the data centers are often incompatible in terms of the temporal and spatial coverage, resolution, origin, format, and map projections. As a result, even when the analysis itself is very simple, considerable time is required to obtain and assemble the data and information into a form ready for analysis. If datasets requested by analysts are not readily available at data centers, the data and information system at the data centers cannot provide the datasets on demand even if the process to make them is very simple. Therefore, analysts have to spend a considerable amount of time ordering and processing the raw data to produce the data they need in the analysis. In a service-oriented science, where highly diversified data and versatile processing functions accessible as services, an "intelligent" mechanism is required to facilitate information discovery and integration over the network and automate the assembly of service chains to provide value-added product.

Two earth science applications are used as examples to help understand the problems that can occur in the Web service based distributed problem solving environment and to illustrate how these problems can be solved using the proposed approach. A short description is given below. The first case is comparatively simple, yet it captures the main features in the Web service-based geospatial knowledge discovery. The second is much more real, complex and difficult, considering that it includes operational data sources from different organizations, various Web services, and the knowledge in the wildfire modeling. The uses cases and related experiments are worked out in Chap. 8.

Use Case 1: Taking a scenario that a user wants to answer the a geospatial question, "what is the landslide risk for location L at time T?". This use case is developed based on the landslide model from the (Sarkar and Kanungo 2004). An OGC-compliant catalogue service can provide help in the search of such an image map with the conditions from the thematic (e.g. landslide susceptibility), spatial (e.g. Dimond Canyon, California, United States) and temporal information (e.g. January 10, 2005). However, an image map of landslide susceptibility is usually a data product generated by an expert analyst. It is not always available and up-to-date for a given region and date. To assess the landslide susceptibility, the expert might have to collect terrain slope data, slope aspect data, land cover data, and vegetation growing condition through the use of Normalized Difference Vegetation Index (NDVI) data. Here, the same problem exists. For example, slope data must be computed from the Digital Elevation Model (DEM) data that is available, while the production of land cover data involves an image classification process. It is possible for the expert to produce all these data products routinely and register them in the catalogue. Yet it is obviously more flexible and intelligent to wrap each computation process as a building block; thus, not only can a high-level data product be produced on demand, but also the flexible composition of these building blocks is possible to satisfy different modeling requirements. Hence, only comparatively low-level data (e.g. DEM or Landsat Enhanced Thematic Mapper (ETM) imagery) need to be updated routinely, which can greatly reduce the cost in data management and maintenance. A service chain can then be created to bind these services and data orderly to generate the landslide susceptibility product for answering this question.

Tables 1.1 and 1.2 list the services and data that can be used to answer this question when introducing Web service as the vehicle for this kind of building blocks. In order to obtain the final answer, these services and data have to be discovered from the catalogue and chained together.

Use Case 2: Assuming a disaster manager, John, wants to know: "What is the possibility of having wildfire(s) in Bakersfield, CA and within a 300 km vicinity tomorrow?" He would go through the following steps to answer the question using the distributed heterogeneous data and various geoprocessing services.

(1) Specifying the metadata description of the desired data product: Through a service registry, John has access to a service registry/catalogue (e.g. CSW) providing descriptions of the available services. There might be several wildfire prediction services available. John knows that earth science applications are always subject to spatial constraints, e.g., a certain wildfire prediction service may be limited to producing wildfire prediction data for a certain place. Thus, John has to first get the bounding box of the area of interest, and

Table 1.1 Services used in this example

Service	Description
Landslide susceptibility	The computational model for landslide susceptibility in this service takes into consideration the factors of terrain slope, terrain aspect, land cover types, and vegetation conditions (through the normalized difference vegetation index, or NDVI) by assigning each a weighting factor and then doing the map algebra computation
Slope	Computes the terrain slope from DEM data
Slope aspect	Generates the terrain aspect from DEM data
ETM NDVI	Calculates ETM NDVI from the near-infrared (NIR) and red bands of ETM images
WICS	OGC web image classification service (WICS) (Yang and Whiteside 2005) which performs the image classification functions (supervised) that can generate the land cover types
WCS	Provides the available geospatial data in the data archives

Table 1.2 Data used in this example

Data	Description
DEM	Terrain elevation data (Dimond Canyon on January 10, 2005[a])
Training image	Label image containing land cover types for the training function in the WICS (Dimond Canyon on January 10, 2005)
ETM image	Image to be classified as land cover types (Dimond Canyon on January 10, 2005)
NIR image	Near-infrared (NIR) band of ETM image for NDVI calculation (Dimond Canyon on January 10, 2005)
Red image	Red band of ETM image for NDVI calculation (Dimond Canyon on January 10, 2005)

[a] To illustrate the solution, the data are assumed to satisfy the temporal condition and registered in catalogue with a long time period

then use it as the filter to get a qualified wildfire prediction service. John knows that a Geocoder service, a Coordinate Transformation Service (CTS) (projecting geographic coordinates to buffer processing coordinate system), a geometry buffer service, a CTS (transforming the projected coordinates to geographic coordinates) and a geometry envelope calculation service can be chained to generate the bounding box for the area with which he is concerned. John uses that service chain first to create the bounding box for the area of concern. In addition, John also specifies the projection for the wildfire prediction image data product: a Lambert Azimuth Equal Area projection (LAMAZ), centered at latitude 45 degree and longitude -100 degree.[1]

(2) Process modeling: John knows that, in order to get a wildfire prediction product for the region within 300 km of Bakersfield, he usually needs to use the output of a buffer process to get the part of the wildfire prediction product of an available, qualified wildfire prediction service that he needs. Thus he must rely on an image cutting service (a service which uses a polygon to cut the image, creating an image containing the values of the desired area only) to create the data product for the area with which he is concerned. Thus, John constructs an abstract process that consists of feeding the output of a buffer process and the wildfire prediction product into an image cutting process.

(3) Creating the executable workflow: John now wants to create an executable service chain that can be stored and routinely create the desired wildfire prediction data product. John finds a wildfire prediction service that, given maximum temperature, minimum temperature, precipitation amount, Leaf Area Index (LAI), Fraction of Photosynthetically Active Radiation (FPAR), land cover/use types (LULC) as input, can generate wildfire prediction data products for California. John searches the catalogue (e.g. CSW) to find the input data for the wildfire prediction service, using the next day's date as the

[1] This is a projection best for the visualization of continental United States.

temporal filter and a bounding box constraint of this wildfire prediction service as the spatial filter. John finds that the National Oceanic & Atmospheric Administration (NOAA) National Digital Forecast Database (NDFD) can provide the weather data and National Aeronautics and Space Administration (NASA) Earth Observing System (EOS) Moderate Resolution Imaging Spectroradiometer (MODIS) products can provide the FPAR, LAI, and LULC.[2] Tables 1.3 and 1.4 show the data and services used.

John needs several general geospatial data processing services to coregister the data sets, the so-called data reduction and transformation services, including data format conversion, coordinate system transformation, and resampling/interpolation/regridding. In some cases, these general services may also be available as optional functions in data request services, such as the WCS. In this example, for the purposes of illustration, the WCS does not provide these optional functions. The operationally available NASA data in the Land Processes Distributed Active Archive Center (LPDAAC) are stored in HDF-EOS data format, and in a sinusoidal grid coordinate reference system at a spatial resolution of 1 km. The MODIS grids are stored as tiles, each covering approximately 1200 by 1200 square kilometers. The operational NDFD data are stored in the GRIB2 data format with a Lambert conformal coordinate reference system and a spatial resolution of 5 km. The fire prediction service takes input data in HDF-EOS format, with LAMAZ projection and 1 km spatial resolution. Preprocessing is needed to transform the NASA and NDFD data into the form that can be readily accepted by the service.

1.3 Research Objectives

The objective of the proposed research in this book is to develop the key technologies for an intelligent geospatial knowledge system based on Web services to

(1) Automate the data discovery and data preprocessing steps in the distributed Web service environment, allowing analysts to focus on the creative process of generating hypotheses and synthesizing knowledge rather than spending large amounts of time on data preparation;
(2) Automate a range of knowledge discovery processes in a limited geospatial domain, using the automated construction and execution of service chains.
(3) Facilitate the construction of complex services and models for geocomputation.

[2] Each FPAR and LAI grid is considered as valid for 7 days, until it is replaced with the next. For prediction purpose, they can be valid on that time. The data are assumed to be valid on that time.

Table 1.3 Services used in this example

Service	Description
Wildfire prediction	OGC WPS process that uses a logistic regression algorithm to provide the computational model for wildfire prediction. It takes into consideration the maximum temperature, minimum temperature, precipitation, leaf area index (LAI), fraction of photosynthetically active radiation (FPAR), land cover/use types (LULC)
Image cutting service (IMCS)	OGC WPS process that uses a polygon to cut an image. The input polygon follows the geography markup language (GML) schema
Geocoder	OGC WPS process that provides the geographic coordinates for the geographic address. The input data and output data follows the OGC OpenGIS Location Service (OpenLS) schema
Buffer	OGC WPS process that performs the spatial operation of buffer. The input and output follows the GML schema
GetEnvelope	OGC WPS process that calculates the bounding box of a feature. The input data and output data follows the GML schema
CTS	OGC WPS process that performs the reprojection computation. It can transform the data from one spatial projection to another spatial projection. The input data and output data follows the OGC WCTS schema
Data format translation service (DFTS)	OGC WPS process that performs the reformating computation. It can transform the data from one file format to another file format
Resolution conversion service (RCS)	OGC WPS process that performs the operations of resampling/interpolation/regridding
OGC CSW	OGC Web-based geospatial catalog service for publication, discovery, and access of geospatial data and services
OGC WCS	Provides the available geospatial data (MODIS and NDFD) in the data archives

Table 1.4 Data used in this example

Data	Description
FPAR	MODIS/aqua fraction of photosynthetically active radiation data product. Operational NASA EOS data (MYD15A2.4), available from the NASA LPDAAC
LAI	MODIS/Aqua Leaf Area Index data product. Operational NASA EOS data (MYD15A2.4), available from the NASA LPDAAC
LULC	MODIS/terra land cover type data product. Operational NASA EOS data (MOD12Q1.4), available from the NASA LPDAAC
Maximum temperature	NOAA NDFD maximum temperature element
Minimum temperature	NOAA NDFD minimum temperature element
Precipitation amount	NOAA NDFD precipitation amount element

The key research issues include,

(1) Developing those interoperable geospatial Web services that can be used to in a distributed environment to discover, request, access, and obtain geospatial data and information;
(2) Intelligently orchestrating interoperable geospatial Web services to generate geospatial process models that can transform data into information and information into knowledge for assistance in making decisions;
(3) Automatically converting geospatial process models to executable service chains that can be invoked and executed on demand;
(4) Management of process models and service chains. The process models and service chains can be archived and catalogued. They can then be advertised as new geospatial services and thus be discovered and used in future geospatial modeling.

By implementing a geospatial knowledge system with advanced standards-based interoperable geospatial Web service technologies, the research addresses many aspects of the technology required for geospatial intelligence analysis and other general applications. The framework will enhance information access and delivery by demonstrating the capability to provide interoperable, on-demand geospatial data access and services tailored to the individual analyst's unique requirements. It also will provide seamless, automated access to data residing in distributed multi-petabyte archives through a common data environment enabled by Web data access interfaces (e.g., WCS, WFS, CSW) compliant with OGC-specifications. The research provides a reference architecture and prototype to demonstrate how interoperable, open systems benefit the end-user community by allowing individuals to contribute reusable geospatial Web service modules that can be dynamically integrated into large web-executable geospatial models and be reused by others in the community as part of a whole system implementation. To this end, the framework accommodates new components (evolvable) and the upgrade or replacement of existing components (maintainable). In the *knowledge management* aspect, the existing process models and the workflow management technology allow analysts to interactively construct new and complex Web-executable geospatial process models. The models could be a data mining, feature extraction, or environmental assessment. The knowledge of the modelers has been captured through the construction of process models and stored in a workflow language. Through a proper peer-review process, the executable models will be kept as virtual geospatial services in the system for sharing. Larger, more complex models can be built upon those existing models. This accumulation of knowledge through sharing and reuse of geospatial process models will allow the system to evolve, increasing its capabilities with time. This community-involved open, accumulated approach for sharing geospatial knowledge by building Web-executable geospatial modules and models will revolutionize the process of geospatial information extraction and knowledge discovery. Therefore, the research will have the profound effects on geospatial knowledge discovery for intelligence analysis.

1.4 Research Approach

To achieve the research objectives, it is necessary to make Web services semantically meaningful, in addition to syntactically expressiveness. For example, the OGC WCS interface unambiguously defined the syntax for requesting a coverage data set. It does not tell, however, how to obtain a surface temperature data set instead of a soil moisture data set. Similarly, the WICS interface defines its input being a multiple-band image, which is essentially is a three dimensional (3D) data array. This input is not different from another service also taking a 3D array as input, such as a color compositing service. Semantic descriptions of Web services and semantic interoperability ensure that the right services are invoked to produce the right outcomes, as opposed to syntactical interoperability, which ensures only that services are invoked using the correct form.

Dynamically and collaboratively sharing and using resources is the concern of the Semantic Web community (Berners-Lee et al. 2001). Semantic Web technologies, which give machine-processable meanings to the documents, allow the semantics of data and services to be used by machines (reasoning) for more effective discovery, integration, and reuse of geospatial data and services. A set of core technologies recommended by the World Wide Web Consortium (W3C) already exists, among them, Resource Description Framework (RDF) (Klyne and Carroll 2004), Web Ontology Language (OWL) (Dean and Schreiber 2004), and SPARQL Protocol and RDF Query Language (SPARQL) (Prud'hommeaux and Seaborne 2006). The Semantic Web community works closely with the Artificial Intelligence (AI) community. Members of the Semantic Web community have applied ontology concepts developed in the AI community to Web Services and search for and manipulation of Web information. Thus, these technologies show considerable promise for better discovery methods by exploiting underlying semantics in the descriptions for geospatial data and services.

With the emergence of Semantic Web, Semantic Web Service has become an area of active research. It is essentially a combination of Semantic Web and Web service technologies, designed to maximize automation and dynamism in all aspects of Web service provision and use, including (but not limited to) discovery, selection, composition, negotiation, invocation, monitoring and recovery (SWSI 2004). The purpose of Semantic Web services is to provide mechanisms for organizing information and services, allowing human queries to be correctly structured for the available application services (the model components and data), thus "automatically" determining the correct relationships between available data and services and build workflows for specific problems. From this point of view, the research described here shares the same goal as Semantic Web services except that this research will deal with geospatial problems in particular. Thus the research will use Semantic Web Service as the vehicle to fulfill the proposed objectives.

To address the first key research issue, a standard-based common data environment is required. The common data environment is a set of standard interfaces for finding and access data in diverse data archives, ranging from small data providers to multiple-petabyte NASA EOS data archives. The environment allows geospatial services and value-added applications to access diverse data provided by different data providers in a standard way without worrying about their internal differences in handling the data. The interface standards for the common data environment are OGC Web Data Services Specifications, including WCS, WFS, WMS, and CSW. Yet current OGC service specification (Percivall 2002) focuses on the syntactical interoperability and does not address the semantic interoperability. This research takes advantage of the emerging Semantic Web technologies to add the richness of semantics, thus provide a promise for the semantic interoperability.

Key research Issues 2–4 are also the concern of automatic service composition, a hot research topic in the general information technology domain. This research address the service composition, the process of creating the service chain, as a procedure including three phases: (1) process modeling, which generates an abstract composite process model consisting of the control flow and data flow among atomic processes; (2) process model instantiation, where the abstract process is instantiated into a concrete workflow or executable service chain; and (3) workflow execution, where the chaining result or workflow is executed in the workflow engine to generate the on-demand data product. Semantic Web technologies have been widely used to enable the automation in the first and second phase, i.e. automatic generation of an executable composite service. Specifically, they are usually combined with the Artificial Intelligence (AI) technologies, especially AI planning methods. The research also uses the AI technologies except the approach deals with geospatial problems. Particularly, the approach is employed into a common service environment which employs a set of standards, developed by W3C, Organization for the Advancement of Structured Information Standards (OASIS) and OGC, to address the service declaration, discovery, binding, invocation and chaining.

1.5 Research Activities

The research activities in this book include the following work:

(1) Literature review, which includes the introduction of Web service technologies, Semantic Web technologies and a survey of automatic service composition.
(2) Ontologies for geospatial data and Web service.

- Incorporate and leverage ontologies of ISO 19100 series (ISO/TC 211, 2007), especially ISO 19115 (ISO/TC211, 2003), and ISO 19136, i.e. GML (Geography Markup Language) (Portele 2007), for the semantic representation of geospatial information.

- Combining general service ontologies from the Semantic Web service area with geospatial domain ontologies to allow semantic representation of geospatial services.

(3) Automated geospatial data and services discovery and access.

- Semantics-enabled discovery of geospatial data and services.
- Seamless, automated access to data residing in distributed archives through a common data environment enabled by Web data access interfaces (e.g., WCS, WFS, WMS, and CSW) compliant with OGC-specifications.
- Automatic access and invocation of standard-based geospatial Web services.

(4) Automatic composition of geospatial Web service

- Domain knowledge-driven automatic construction of process model. Different planning methods are evaluated by cases.
- The role of metadata tracking in the geospatial service composition.
- Automated geospatial Web service chaining, binding, and execution based on the process model.

(5) Implementation of a prototypical system with applications that can demonstrate that such semantics-enabled intelligent geospatial Web service can maximize the potential of individual data and services and significantly advance the geospatial knowledge discovery.

1.6 Reader's Guide

An outline of the structure of this book is shown as follows:

The first part of the book serves as the foundations for the research work. Chapters 2, 3 and 4 provide a general introduction to the fundamental concepts to the work. However, they do not attempt to give a thorough introduction to these concepts, but the incorporated references can serve as the hints for interested readers to learn more about these concepts.

Chapter 1 provides a general introduction and characterization of the book, including research context, motivations, use cases, research objectives, research approach, research activities, and book outline.

Chapter 2 recapitulates Web service and OGC Web service technologies relevant to the book, in particular the fundamentals for the interoperability and integration of Web services, the common data environment and common service environment in this research.

Chapter 3 briefly introduces the concept of Semantic Web. Furthermore, it introduces Semantic Web service technologies and current development towards Geospatial Semantic Web.

Chapter 4 provides an overview of literatures on Web service composition. In particular, some basic AI planning concepts and methodologies relevant to the approach are introduced.

The second part of the book presents the approaches towards the semantics-enabled intelligent geospatial Web service.

Chapter 5 introduces the usage of ontologies for representing semantics of geospatial data and Web service.

Chapter 6 presents semantics-enabled discovery of geospatial data and services. A survey of current literature on semantics-enabled service registry is provided. It shows how semantic search is supported in an ebRIM profile based CS/W.

Chapter 7 addresses different planning approaches for automatic service composition. In particular, it illustrates their application, benefits, limitations and relations through case studies.

The third part of the book (Chap. 8) presents the implementation and evaluation of the proposed approaches.

Finally, Chap. 9 summarizes work and presents future directions.

Chapter 2
Geospatial Web Service

2.1 Interoperability of Web Service

Interoperability is the capability to exchange information, execute programs, or transfer data among various functional units in a manner that requires the user to have little or no knowledge of the unique characteristics of those units (Percivall 2002). There are two levels of interoperability[1]: syntactical interoperability and semantic interoperability (Percivall 2002). The former requires that there is a technical connection, i.e., that the data can be transferred between Web services. It does not provide an interpretation for the content transferred in the connection. The latter assures that the contents of data and services are correctly understood when data/services are connected.

Syntactical interoperability of Web services is achieved mainly using two common Web service standards: Web Services Description Language (WSDL) (Christensen et al. 2001) and Simple Object Access Protocol (SOAP) (W3C 2007a). WSDL is used to describe a Web service in terms of its interfaces and SOAP formalizes the XML (Extensible Markup Language)-based message transportation between Web services. In the geospatial community, OGC has defined a series of interface specifications for the interoperability of geospatial Web services, e.g. WFS, WMS, WCS, WCTS, WICS and WPS. These specifications follow the principles for geospatial Web services defined in ISO 19119 (ISO/TC 211 2005), and describe the structure of content transferred between Web services. For example, WCS defines the standard interface and message type for Web services providing coverage data, yet it does not formalize the conceptualization of content.

To achieve semantic interoperability, the conceptualization of content should be expressed formally and explicitly. This can be achieved by using ontologies. An

[1] Some may argue the structural interoperability, e.g. mapping the elements in the output message structure of one service to the input message structure of the next dependable service. This book follows the definition of syntactical and semantic interoperability from the OGC Abstract Service architecture. It treats this kind of structure difference as the issue related to the semantic interoperability.

P. Yue, *Semantic Web-based Intelligent Geospatial Web Services*, SpringerBriefs in Computer Science, DOI: 10.1007/978-1-4614-6809-7_2,

ontology is a formal, explicit specification of a conceptualization that provides a common vocabulary for a knowledge domain and defines the meaning of the terms and the relations between them (Gruber 1993). Ontologies are crucial to making the semantics of the exchanged content machine-understandable. OWL is recommended by W3C as the standard Web ontology language. It is designed to enable the creation of ontologies and the instantiation of these ontologies in the description of resources. The work reported in this book addresses the semantic interoperability through the introduction and design of OWL-based ontologies conveying semantic information on geospatial services and data.

2.2 Integration of Web Services

Currently, there are many individual standalone services available over the Internet. However, it is impossible for individual standalone services to meet all service requirements of many users. Such information requests could be met by dynamically chaining multiple services provided by single and multiple service providers. The service-oriented architecture (SOA) recognizes this and tries to construct a distributed, dynamic, flexible, and re-configurable service system over Internet that can meet many different users' information requirements. It provides the basis for the integration of Web services. There are three key actors in SOA (Fig. 2.1): requestor, provider and broker. The requestor is the user who requires the information services. The provider is the standards-based individual service. The broker is a meta-information repository (e.g., a registry, catalog or clearinghouse). The interactions among these actors involve the operations of publishing, finding and binding. Service composition introduces a new operation into SOA, chaining, which combines services into a dependent series to accomplish a larger task. SOA is the basis for automatic service composition, since services management functions such as registration, discovery, accessing, and execution are well positioned under this structure and these functions are the basic units in the whole automation process.

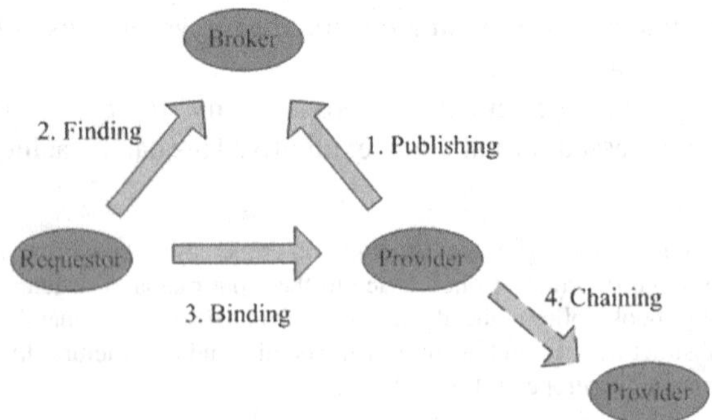

Fig. 2.1 The basic SOA operations

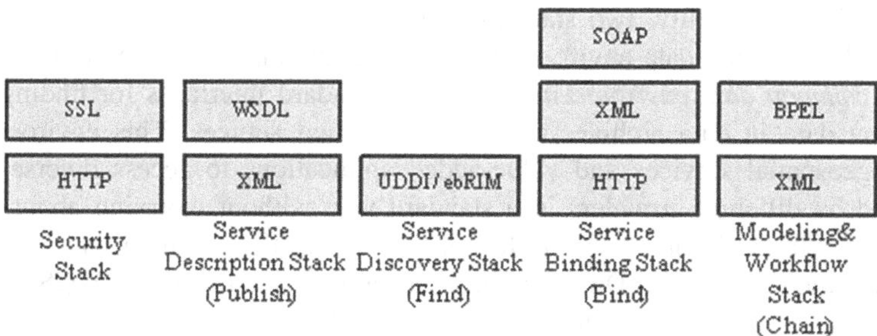

Fig. 2.2 The major web service standards

The Web service technologies follow the publish-find-bind paradigm in Service-Oriented Architecture (SOA) and have service discovery, description, and binding layers (Papazoglou 2003).

In order for SOA to work, various standards related to all aspects of service operations are needed. The major international bodies setting the Web service standards are W3C and OASIS. The major standards related to services are shown in Fig. 2.2.

2.3 Geospatial Web Service

In the geospatial Web services area, OGC is the major organization working on developing geospatial Web services standards by adapting or extending the common Web service standards. Through the OGC Web Services (OWS) testbeds, OGC has been developing a series of interface specifications under the OGC Abstract Service Architecture (Percivall 2002), including WFS, WMS, WCS, CSW and WPS. Currently, OGC Web services are not equivalent to the W3C SOAP-based Web services. Most of OGC Web service implementations provide access via HTTP GET, HTTP POST and do not support SOAP. The registry service, i.e. CSW, provides the discovery not only on the services, but also on the geospatial data. An Electronic Business Registry Information Model (ebRIM) profile of Catalogue Services for the Web (Martell 2008) has been developed and recommended for CSW implementation. Conceptually, the OWS also follows the publish-find-bind paradigm in the SOA and has service discovery, description, and binding layers corresponding to UDDI (Universal Discovery Description and Integration) (OASIS 2004), WSDL, and SOAP in the W3C architecture. In addition, OGC is also attempting to integrate the W3C Web services standards into the OWS framework by providing WSDL descriptions for OGC Web services.

As identified by Di (2005a), a framework for intelligent geospatial knowledge systems requires interoperability of both geospatial data and services in order for the system to be able to pull out and chain data and services from providers to complete user requests for geospatial information and knowledge. In order to facilitate interoperability, two standards-based interoperability environments are needed: the common data environment and the common service environment.

The *common data environment* is a set of standard interfaces for finding and accessing data in data archives of varied sizes and sources. This environment allows geospatial services and value-added applications to access diverse data provided by different providers in a standard way without worrying about their internal handling of data. The interface standards for the common data environment are the OGC Web Data Services Specifications, including WCS, WFS, WMS, and CSW. And the data-encoding standard is GML, which is well developed to describe geometries and geographical relations.

The *common service environment* is a set of standard interfaces for service declaration, description, discovery, binding, chaining, and execution. This environment allows geospatial knowledge systems dynamically to generate user-specific geospatial information/knowledge by discovering and chaining standards-compliant services supplied by service providers. The requirements for this set of standards in a geospatial knowledge system are very similar to the requirements in mainstream Web services technology. Therefore, the standards used in the mainstream Web service arena can be adopted for geospatial knowledge systems.

This work will use the OGC standards for the data finding and access, OGC and W3C standards for the Web services. OGC specifications are widely used by geospatial communities for sharing data and resources and are becoming ISO standards. For standards that are not available at OGC, the research will adopt W3C Web services standards because the system to be developed is basically the geospatial version of W3C Web service system.

Chapter 3
Geospatial Semantic Web

3.1 Semantic Web Architecture

Inspired by Tim Berners-Lee (Berners-Lee 1998), inventor of the Web, a growing number of individuals and groups from academia and industry have been evolving the Web into another level—the Semantic Web. By representing not only words, but their definitions and contexts, the Semantic Web provides a common interoperable framework in which information is given a well-defined meaning such that data and applications can be used by machines (reasoning) for more effective discovery, automation, integration and reuse across various application, enterprise and community boundaries. Compared to the conventional Web, the Semantic Web excels in two aspects (W3C 2001): (1) common formats for data interchange (the original Web only had interchange of documents) and (2) a language for recording how the data relates to real world objects. With such advancements, reasoning engines and Web-crawling agents can go one step further—and inductively respond to questions such as "*which airfields within 500 miles of Kandahar support C5A aircraft?*" rather than simply returning Web pages that contain the text "airfield" and "Kandahar", which most engines do today.

Figure 3.1 shows the hierarchical architecture of the Semantic Web. At the bottom level, XML provides syntax to represent structured documents with a user-defined vocabulary but does not necessarily guarantee well-defined semantic constraints on these documents. And XML schema defines the structure of an XML document. RDF is a basic data model that identifies objects ("resources") and their relations to allow information to be exchanged between applications without loss of meaning. It is based on a graph model composed of triples. RDFS (RDF Schema) is a semantic extension of RDF for describing the properties of generalization-hierarchies and classes of RDF resources. OWL adds vocabulary to explicitly represent the meaning of terms and their relationships, such as relations between classes (e.g. disjointness), cardinality (e.g., "exactly one"), equality and enumerated classes. The logic layer represents the facts and derives knowledge, and deductive process and proof validation are deduced by the proof layer. A digital signature can be used to sign and export the derived knowledge. A trust

P. Yue, *Semantic Web-based Intelligent Geospatial Web Services*,
SpringerBriefs in Computer Science, DOI: 10.1007/978-1-4614-6809-7_3,

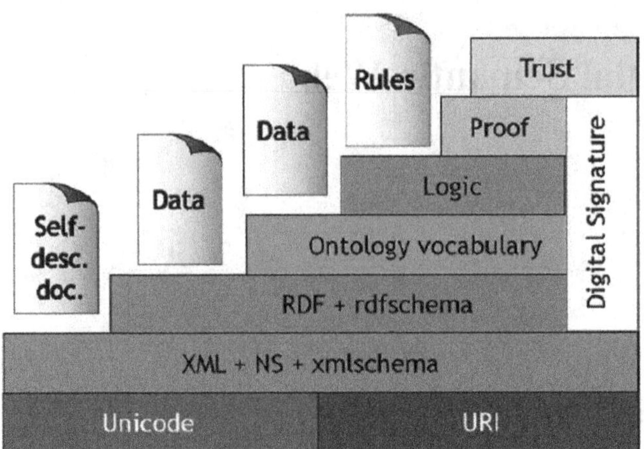

Fig. 3.1 Semantic web architecture (Berners-Lee 2000a)

layer provides the trust level or a rating of its quality in order to help users building confidence in the process and quality of information (Antoniou and Harmelen, 2004). Currently there is no consensus on how a rule layer could look like and some proposals exist such as Rule Interchange Format (RIF) (Kifer 2008), RuleML (Boley et al. 2001), Notation3 (N3) (Berners-Lee 2000b), and SWRL (Horrocks 2004). The top layers providing proof and trust are starting to be addressed by research today. This book focuses on the approach from the ontology layer.

3.2 Semantic Web Service

With the advancement of Semantic Web, there are a number of representative technologies concerning frameworks for semantics in Web Services. Each of them has its own key concerns and emphasis. Web service ontology (OWL-S) (Martin et al. 2004), which is based on OWL, is primarily concerned with service composition. It models individual services as atomic processes with corresponding operation/functionality, input/output, pre/post-conditions. Service chain is represented as the composite process with control constructs defined based on the workflow pattern such as sequence, parallel split, and choice. Web Service Modeling Ontology (WSMO) (Bruijn et al. 2005) is defined based on the Web Service Modeling Framework (WSMF) (Fensel and Bussler 2002) which consists of four different main elements for describing Semantic Web services: (1) ontologies that provide the terminology used by other elements (2) goals that state the intentions that should be solved by Web services (3) Web service descriptions that define various aspects of a Web service, and (4) mediators which resolve interoperability problems. A more expressive logical language, named Web Service

Modeling Language (WSML), is used as the basis language framework for WSMO, compared with OWL for OWL-S. Semantic Web Services Framework (SWSF) (Battle et al. 2005) is proposed by the Semantic Web Services Initiative (SWSI) and intended to serve as a theoretical and comprehensive framework for semantic specifications of Web services. It also consists of a representational language, called Semantic Web Services Language (SWSL), and an ontology, called Semantic Web Services Ontology (SWSO). Based on the two sublanguages in SWSL, first-order logic (FOL) based SWSL (SWSL-FOL) and logic programming (LP) based SWSL (SWSL-Rules), SWSO has two corresponding types of ontologies, including FLOWS (First-Order Logic Ontology for Web Services) and ROWS (Rules Ontology for Web Services). SWSO also models the Web services as processes, the same as OWL-S does. Yet the process ontology is built on a mature preexisting ontology of process modeling concepts, the Process Specification Language (PSL). Apart from defining a service ontology, Web Service Semantics (WSDL-S) (Akkiraju et al. 2005) and Semantic Annotations for WSDL (SAWSDL) (Farrell and Lausen 2006) aim to extend existing WSDL elements with semantic annotations. With the emergence of the RESTful services (Pautasso et al. 2008), a similar idea to SAWSDL is adopted by a W3C Member Submission called Semantic Annotations for REST (SA-REST) (Gomadam et al. 2010). These methods provide a practical way with less effort to describe the semantics of Web service within the legacy of current Web service standards.

3.3 Geospatial Semantic Web

Parallel to the development of the Semantic Web, Geospatial Semantic Web—a geospatial domain-specific version of the Semantic Web, is initiated. Because geospatial information is heterogeneous, i.e. multi-source, multi-format, multi-scale, and multi-disciplinary, the importance of semantics on accessing and integration of distributed geospatial information has long been recognized (Sheth 1999). The advent of the Semantic Web promises a generic framework to use ontologies to capture the meanings and relations for information retrieval. But this framework does not relate explicitly to some of the most basic geospatial entities, properties and relationships that are most critical to a particular geospatial information processing task. To better support the discovery, retrieval and consumption of geospatial information, the Geospatial Semantic Web is initiated to create and manage geospatial ontologies to capture the semantic network of geospatial world and allow intelligent applications to take advantage of build-in geospatial reasoning capabilities for deriving knowledge. It will do so by incorporating geospatial data semantics and exploiting the semantics of both the processing of geospatial relationships and the description of tightly-coupled service content (Egenhofer 2002; Lieberman et al. 2005). The Geospatial Semantic Web was identified as an immediately-considered research priority early in 2002 (Fonseca and Sheth 2002) by University Consortium for Geospatial Information Science

(UCGIS). Since 2005, OGC has issued the Geospatial Semantic Web Interoperability Experiment (GSW IE) aiming to develop a method of discovering, querying and collecting geospatial content on the basis of formal semantic specifications (Kolas et al. 2005, 2006; Kammersell and Dean 2006; Lutz and Kolas 2007). In this experiment, five types of ontologies are identified, including base geospatial ontology, feature data source ontology, geospatial service ontology, geospatial filter ontology and domain ontology. Based on these ontologies, a user's query can be translated to the data source semantic queries via semantic rules, and then transformed to WFS query through Extensible Stylesheet Language Transformations (XSLT) (Clark 1999). The query is represented using the SPARQL, and the semantic rules are represented using SWRL. More recently, GeoSPARQL, a spatial extension of SPARQL for querying spatial RDF data (Battle and Kolas 2012), has been standardized in OGC. The standard also provides capabilities to exploit the increasing amount of geospatial data in the Semantic Web published using the Linked Data approach (Bizer et al. 2008).

Chapter 4
Automatic Service Composition

Broadly speaking, service composition, the process of creating a service chain, can address many aspects of Web service provision and use, including discovery, selection, composition, negotiation, invocation. However, following the main efforts published in the literature, the automatic service composition issue addressed here focuses on the methods for dynamic service discovery and composition in automatic generation of composite service. This chapter briefly sketches the methods from the business perspective and AI research area. In particular, some basic concepts of workflow and AI planning are introduced. Finally, related work in geospatial domain is introduced.

4.1 Business Perspective

The Workflow Management Coalition (WfMC) is s a non-profit, global organization of adopters, developers, consultants, analysts and university/research groups engaged in Business Process Management (BPM). The WfMC has been responsible for the creation of a workflow reference model and a glossary of standardized workflow terminology. The WfMC Terminology and Glossary document (WfMC 1999) defines the *Workflow* as follows: the automation of a business process, in whole or part, during which documents, information or tasks are passed from one participant to another for action, according to a set of procedural rules. Workflow management is concerned with the declarative defnition, enactment, administration, and monitoring of business process. A business process consists of activities related by data and control flow relationship. An activity is typically performed by executing a program, enacting a human/machine action, or invoking another process. It concerns about the order of (atomic) activities, while data flow focuses on the data exchange among the activities.

Some research on the service composition has been performed from a business perspective focusing on workflow-based composition. Workflow is a key technology for automating business process that involve access to several applications.

SpringerBriefs in Computer Science, DOI: 10.1007/978-1-4614-6809-7_4,

However, traditional workflow systems are ineffective when considering the needs of Web-based applications, with their complex partnerships, possibly qmong a large number of highly evolving process. There are already some Web service composition languages such as the Web Services Business Process Execution Language (WSBPEL, shortly known as BPEL) (OASIS 2007), the XML Process Definition Language (XPDL) (WfMC 2008), and the Yet Another Workflow Language (YAWL) (Aalst and Hofstede 2004). The control-flow aspect of such languages is comparable to that developed in workflow research (Aalst 2003) with similar flow control constructs such as sequence, and split. In addition, workflow-based systems are making efforts to support composite Web services (Benatallah et al. 2001; Casati et al. 2001). Typical examples are eFlow (Casati et al. 2000) and e-speak (Casati et al. 2001) developed by Hewlett-Packard laboratories. The eFlow platform is a workflow platform for specifying, enacting, and monitoring composite services while e-speak supports composite Web services. Composite services are modeled as business processes, enacted by a service process engine. A composite service is modeled by a graph that defines the order of execution among the nodes in the process. Nodes can be bound automatically with concrete services. Thus automation focuses mainly on the automation of the instantiation process. Business efforts focus mainly on defining standards for composing Web services (Aissi et al. 2002) (e.g., BPEL) and providing platforms to enable B2B interaction on the Web (e.g., IBM WebSphere) (Medjahed et al. 2003). The conceptual level of composition (process model) is usually designed manually. Since BPEL provides rich vocabulary and control structure, and is widely supported by commercial vendors and open-source communities, it is becoming the de-facto standard for describing the control logic required to coordinate those Web services participating in a workflow (Akram et al. 2006; Friis-Christensen et al. 2009).

4.2 AI Planning

There is already significant literature addressing the problem of automatic service composition through AI planning. Russel and Norvig (2003) define planning as follows: “The task of coming up with a sequence of actions that will achieve a goal is called planning”. In planning, there are three types of representational entities: states, goals and actions. The world or a specified domain is modeled as a set of states that can be divided into initial states and goal states. Goals are partially specified states that can be achieved through actions from the initial states of the world. An action is specified in terms of the preconditions and the effects (post-conditions). The preconditions are the states that must hold before the action can be executed, and the effects are the state changes when the action is executed. Thus, the assumption for Web service composition as a planning problem is that a Web service can be specified as an action. As a software component, a Web service takes input data and produces output data. Thus, the input and the output parameters can be treated as the preconditions and effects respectively. Furthermore, the Web service might alter the state of the world after its execution. Then, the world state before service execution is the precondition, and the new state generated after execution is the effect (Rao and Su 2004).

The semantics for inputs, outputs, preconditions and effects (i.e. IOPE semantics) addressed in the Semantic Web Service technologies are widely used in most AI planning methods for automatic service composition. Most AI planning methods use OWL-S for their service model representation (Ponnekanti and Fox 2002; Sirin et al. 2004; Klusch et al. 2005). The OWL-S descriptions are often transformed into a planning problem, while semantic information is used for the enhancement of the composition process (Hatzi et al. 2012). Peer (2005) summarizes the basic planning paradigms and knowledge-oriented paradigms in AI planning. A common characteristic of these methods is that they are subject to constraints and assumptions that limit their use for wide applications. Special efforts need to be performed for a particular application problem. The logical representation of services plays an important role in these methods.

4.3 Geospatial Domain

There are already some geoscience efforts for geospatial Web service composition. Di et al. (2005b) introduce a framework for automatic Geospatial Web service composition. OWL-S is adopted as an experimental representation of a geospatial Web service. The other is Geosciences Network (GEON) (Jaeger et al. 2005). Geospatial Web services, including data (GML representation) provider services and customized services with vector data processing functionalities, are sampled to compose a workflow manually in the KEPLER system (Ludäscher et al. 2005), which provides a framework for workflow support in the scientific disciplines. The major feature of the KEPLER system is that it provides high-level workflow design while at the same time hiding the underlying complexity of technologies as much as possible from the user. Both Web service technologies and Grid technologies are wrapped as extensions in the system. For example, individual workflow components (e.g., data movement, database querying, job scheduling, remote execution) are abstracted into a set of generic, reusable tasks in a grid environment (Altintas et al. 2004). Thus, combining a knowledge representation technique (e.g., OWL and OWL-S), with the lower level generic/common scientific workflow tasks in the KEPLER system, is a worthwhile technique for attempting to minimize or eliminate human intervention in the generation and instantiation of workflow. OWL is introduced into SEEK (a similar and major contributor to KEPLER) to enable automatic structural data transformation in the data flow among services. The transformation is based on ontology and registration mapping of input and output structural types to their corresponding semantic types (Bowers and Ludäscher 2004).

More related to the service and data discovery are efforts to add semantically augmented metadata information to annotate data and services (Lutz and Klien 2006; Maue et al. 2012). Ontologies, related in both simple taxonomic and non-taxonomic ways, are employed using subsumption reasoning (Baader and Nutt 2003) to improve service discovery and the recall and resolution of data. Template operations are introduced for semantic annotation of services input/output and functionality (Lutz 2004). WSMO is used to facilitate discovery and invocation of semantically described geospatial Web Services (Roman et al. 2006; Zaharia et al. 2009). Lemmens et al. (2006) experimented with WSDL-S in their use-case implementation.

The implementation of services and service chains is not limited to the standards and technologies from OGC and W3C used in this book. OGC Web services, W3C SOAP-based Web services, and RESTful services are available for implementation. Some efforts have been devoted to make them work together, such as defining WSDL for OGC services (Sonnet 2005), and using WSDL 2.0 as the bridge between REST and W3C Web service (W3C 2007b; Lucchi et al. 2008). In addition to the BPEL-based service chaining approach, there is an OGC WPS approach for Web Service Orchestration (Stollberg and Zipf 2007). However, a comparative analysis shows that the BPEL-based implementation is more mature (Friis-Christensen et al. 2009).

4.4 Summary

The related work described so far helps identify the particular requirements of the geospatial domain that automatic service composition satisfies.

- Data-intensive: Geospatial Data for processing is always high volume and diversified with inherent disciplinary complexity. Data plays an important role in geospatial service composition since its rich, explicit, and formalized semantics (other than traditional metadata) of geospatial data allow a machine to understand and automatically discover the appropriate data (other than by keyword matching) for a service's input. Formal conceptualization of data semantics requires the combination of geospatial domain knowledge with certain knowledge representation techniques, e.g., formal ontologies. Semantic Web standards such as OWL provide such support. In addition, the complete data semantics can help the metadata tracking in the service chaining which ensures the trustable data for users (Alameh 2003).
- Compute-intensive: Geoprocessing functions are complex, time-consuming and data-dependent. For compute-intensive applications, offline planning is preferred to online planning. In offline planning, the process model for service composition is generated before the execution of the service component, e.g., SWORD (Ponnekanti and Fox 2002). Online planning is useful usually when the information for the generation of the process model is incomplete and thus

requires the invocation of a service component as the information provider. The actual process model is created at run time, e.g., SHOP2 (Sirin et al. 2004). Given the resources consumed by geospatial processing services, offline planning can bring predictability and efficiency. Alternative process models should be created to deal with the possible inapplicability of certain process models. In addition, service semantics also need to be explicitly formalized and the inherent relation to data should be identified. The ontology descriptions using Semantic Web Service technologies will enable the reasoning and chaining of services such as the aggregate service and workflow managed chaining identified in the OGC geoprocessing architecture.

- Analysis-intensive: Geospatial application involves diverse sources of data and complex processing functions. Analysis-intensive applications require that the inherent relations between multiple geospatial data and services should be captured at an upper level. These relations can be constructed through geospatial ontologies and rules, serving as the knowledge base for AI methods. One example rule is that WCTS can be introduced in the service chain automatically when the spatial projection of the available data can not satisfy the spatial projection requirement of a service's input.

Chapter 5
Semantics for Geospatial Data and Services

5.1 Ontology Approach

Ontologies have been used in the geospatial domain for information integration and semantic interoperability. By mapping concepts in a geospatial Web resource (e.g. geospatial data, service, or geoprocessing service chain) to ontological concepts in the geospatial domain, the semantics of that geospatial resource can be explicitly defined. To provide the semantic concepts, the research uses ontologies represented using Semantic Web technologies. OWL is used as the basic technology. The foundation of knowledge representation formalism for OWL is the description logic (DL) (Baader and Nutt 2003). DL is more like an object-oriented approach to knowledge representation. The basic elements of description logics are *concepts*, *roles*, and *constants*. In the Web ontology context, they are also commonly named *classes*, *properties*, and *individuals* respectively. Concepts group individuals into categories, roles stand for binary relations of those individuals and constants stand for individuals.

The expressive power of different DL languages is subject to the set of constructors and axioms in each language. Generally, the particular selection of constructors and axioms is made so that inference procedure is decidable. Constructors are a set of symbols formalized for the definition of concepts and roles. There are two types of constructors: concept-forming constructors and role-forming constructors. These constructors can be used to construct complex concepts and roles from atomic concepts and atomic roles.

Example 5.1.1 (HDFEOS) Given the atomic class MD_Format and the atomic property name_MD_Format, we can describe a GeoTIFF File Format using constructors as

MD_Format $\sqcap$ $\exists$name_MD_Format.application/GeoTIFF

A DL knowledge base (KB) comprises two components: *TBOX* and *ABOX*. TBox consists of a set of terminological axioms which make statements about how concepts or roles are related to each other. ABOX introduces individuals, i.e.

SpringerBriefs in Computer Science, DOI: 10.1007/978-1-4614-6809-7_5,

instances of a class, into the knowledge base and asserts the properties of these individuals.

There are two types of reasoning in DL: TBOX reasoning and ABOX reasoning. In TBOX reasoning, a basic type of reasoning is to determine whether or not a concept is subsumed by another concept (i.e. subsumption reasoning). For example, in a geospatial ontology represented using OWL, if some "subClassOf" axioms are added to signify that "NDVI" is a sub-category of "Vegetation_Index" and "ETM_NDVI" is a sub-category of "NDVI", then DL reasoners can determine that "ETM_NDVI" is subsumed by "Vegetation_Index" using subsumption reasoning. The other type of reasoning, ABOX (Assertional Box) reasoning, is to determine whether a particular individual is an instance of a given concept description, or relations between individuals. For example, if a class "GeoTIFF" is defined to be a subclass of "MD_Format" with the only restriction that the inherited property "name_MD_Format" has a string value "application/GeoTIFF", DL reasoners can use ABOX reasoning to determine whether a particular individual of "MD_Format" is an instance of the class "GeoTIFF".

In practice, according the generality of concepts, the DL knowledge base can be organized hierarchically with a special tree-like data structure, called taxonomy. New facts can be added to a taxonomy through an efficient classification process. This taxonomy allows queries to be answered efficiently and thus makes it practical to consider extremely large knowledge bases.

5.2 Geospatial Semantics in SOA

Geospatial semantics are those that convey content information about geospatial data, entities, phenomena, functionalities, relationships, processes, services, etc. The scope of geospatial semantics can be extremely broad. A number of research projects have been started in this subject, e.g., SWEET (Raskin and Pan 2005). This book focuses on defining data and service semantics that enable dynamic and automatic composition of geospatial Web service chains to achieve a complex geospatial goal that involves heterogeneous data and multiple services. In order to establish geospatial semantics, the semantics of Web service must first be understood. In the Web service domain, semantics can be classified into four types (Sheth 2003): (1) data/information semantics, (2) functional/operational semantics, (3) execution semantics, and (4) Quality of Service (QoS) semantics.

Data semantics annotate the semantics of input and output data in a Web service operation. Functional semantics represent the semantics for a service function. Execution semantics specify the requirements of a service such as the pre-conditions and post-conditions/effects. QoS semantics provide the quality criteria for service selection. For example, a service that calculates the terrain slope from DEM data may require the HDF-EOS data format as a precondition and DEM data as input. It generates the slope as output. The functional semantics for this service can be represented by using the slope entity class in an ontology called Functional

Ontology, in which each concept/class represents a well-defined functionality (Cardoso and Sheth 2005).

As mentioned in Sect. 3.2, current researches on the Semantic Web Service technologies provide the choice of OWL-S, WSMO, WSDL-S, SAWSDL, and SWSF. WSMO and SWSF do not limit their knowledge representation to description logic. Thus, their definitions are not built upon OWL as OWL-S is. WSDL-S and SAWSDL aim to extend existing WSDL elements with semantic annotations; thus, they are not defining a complete ontology framework for Web services as OWL-S does. Most previous work uses OWL-S, and many tools are available. OWL-S can be selected as the starting point for the semantic description of geospatial Web services. OWL-S also provides a "Composite Process" ontology that contains the control and data flow among subprocesses. The control flow specifies the ordering and conditional execution of subprocesses, while the data flow focuses on data exchange among the subprocesses. Therefore, OWL-S can be used to describe the semantics of geoprocessing service chains.

This research primarily focuses on automatic service composition based on geospatial data, functional, and execution semantics, leaving the QoS semantics oriented composition to future work. The subsequent sections show how geospatial DataType and ServiceType ontologies are designed and how they are incorporated into SOA for services integration.

5.2.1 Geospatial DataType and ServiceType Ontologies

Geospatial DataType ontology conceptualizes scientific meanings of distributed geospatial data, thus it can be used to annotate the semantics of input and output data in a geospatial service operation. Furthermore, the DataType ontology can be enriched with metadata ontologies to allow more precise description of geospatial data, and support cross-metadata-standards discovery (Bermudez 2004) of geospatial data through additional semantic relations (e.g., "disjoint" and "equivalent") among terms in different metadata standards such as ISO 19115 and the FGDC metadata standard. An example of such ontology derived from the conceptualization of the Global Change Master Directory (GCMD) (Olsen et al. 2004) science keywords is shown in Fig. 5.1. The figure was captured using Protégé (http://protege.stanford.edu/), a freely available tool that can support OWL. Geospatial ServiceType ontologies are defined according to the scientific problems that geospatial services focus on solving. GCMD provides a comprehensive hierarchical keyword list for services, which can be conceptualized into geospatial ServiceType ontology (Fig. 5.2).

The entity classes in geospatial DataType and ServiceType ontologies describes which entities can possibly exist in the geospatial domain, which in turn are used to represent the data and functional semantics in the geospatial Web service (Fig. 5.3). In addition, it provides the RDF structure (see Sect. 5.2.5) for the XSLT

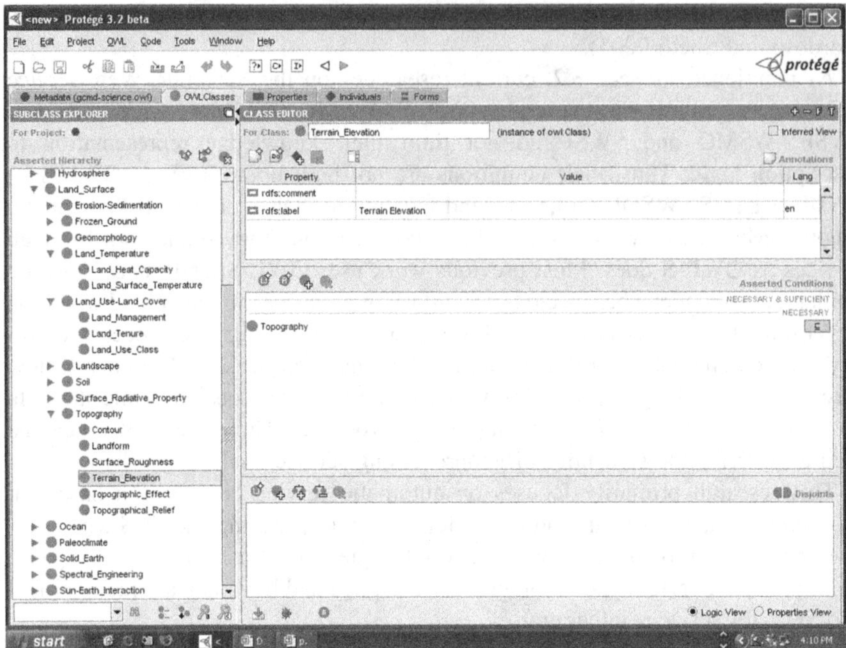

Fig. 5.1 Geospatial DataType ontology based on the GCMD science theme

in the service grounding of OWL-S. In these aspects, they can be treated as the conceptual schema for semantic annotations of geospatial Web services.

5.2.2 Geospatial Semantics for Providers

In SOA, service providers supply services over the Internet. As mentioned before, OWL-S is used to describe the semantics of geospatial Web services. An Unified Modeling Language (UML) graph (Fig. 5.4) is provided to help illustrate how to describe a WCS using OWL-S.

OWL-S is structured in three main parts: (1) service profile: what a service does (advertisement), e.g., "WCS" as a "ServiceType" and "Coverage" as an output "DataType" in Fig. 5.3. (2) service model: how a service works (detailed description), e.g., a series of input parameters which are identified in the service model. (3) service grounding: how to assess a service (execution), e.g., the output "DataType" of WCS is grounded to the output message of the GetCoverage operation defined in the WCS WSDL using an XSLT transformation. The service profile and service model concern the semantic description of the Web service using the geospatial datatype and servicetype ontologies. The service grounding

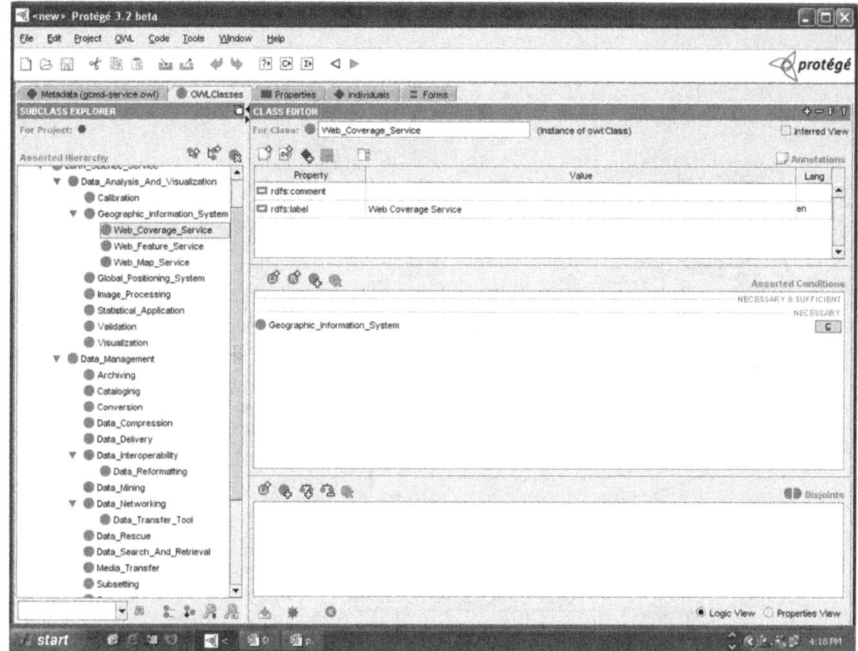

Fig. 5.2 Geospatial ServiceType ontology based on the GCMD service types

describes the relation of the semantic description to the syntactic description of a service.

Table 5.1 shows a snippet of WSDL and OWL-S for the slope computation service. Geospatial DataType (e.g., Terrain_Elevation) and ServiceType (e.g., Slope) are linked into the OWL-S descriptions. The service grounding part of OWL-S provides information on how to bridge the syntactic and semantic worlds, e.g., grounding the input/output ontology concepts to the input/output message of WSDL using XSLT (Table 5.1).

One of the major efforts in the service grounding is to focus on the specification of the XSLT transformation between service messages and OWL-S parameters, since the ontology entity's RDF structure is not always consistent with the grounding message structure[1] of individual services. Two types of elements in the message structure should be differentiated in the grounding description: the elements whose values are passed along in service chains and those whose values are not passed along. Table 5.2 shows some examples. The "service" element in the WICS[2] GetClassification message does not get a value from its precedent services.

---

1 Most geospatial Web services provide access via HTTP GET, HTTP POST, and SOAP which can be described through WSDL interface. Thus WSDL grounding of OWL-S is discussed in this work.

2 WICS version 0.0.20.

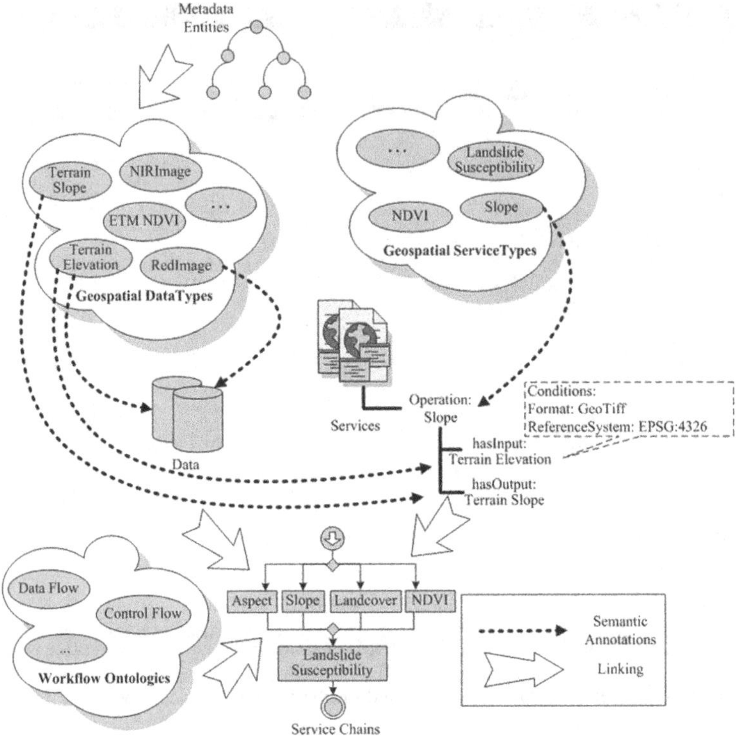

Fig. 5.3 Semantic descriptions for geospatial data, services and geoprocessing service chains

Therefore, to enable automation, this work sets the grounding information for the "service" element of WICS with hard-coded text "WICS", while the "sourceURL" and "sourceFormat" elements can get values at runtime from the RDF structure of the "Data Type" output in the precedent service WCS.

In the past several years, OGC has made significant progresses on the standardization of geospatial Web services. Since the geospatial applications include both OGC-compliant and non-OGC-compliant Web services, this work have developed two groups of OWL-S descriptions for the two categories of geospatial Web services. The OWL-S descriptions for OGC-compliant Web services focus on the semantic representation of the standard interfaces and messages. It is possible to define some common OWL-S grounding representation for all OGC service instances under the same standard interface and message with the premise of the same semantics. For example, different WCS service instances can share the common XSLT transformation information (example in Table 5.2) in service grounding.

Although the OGC Service Architecture abstract specification has listed a series of geographic services that could be standardized (Percivall 2002), standard interface protocols are currently defined for only a very limited number of

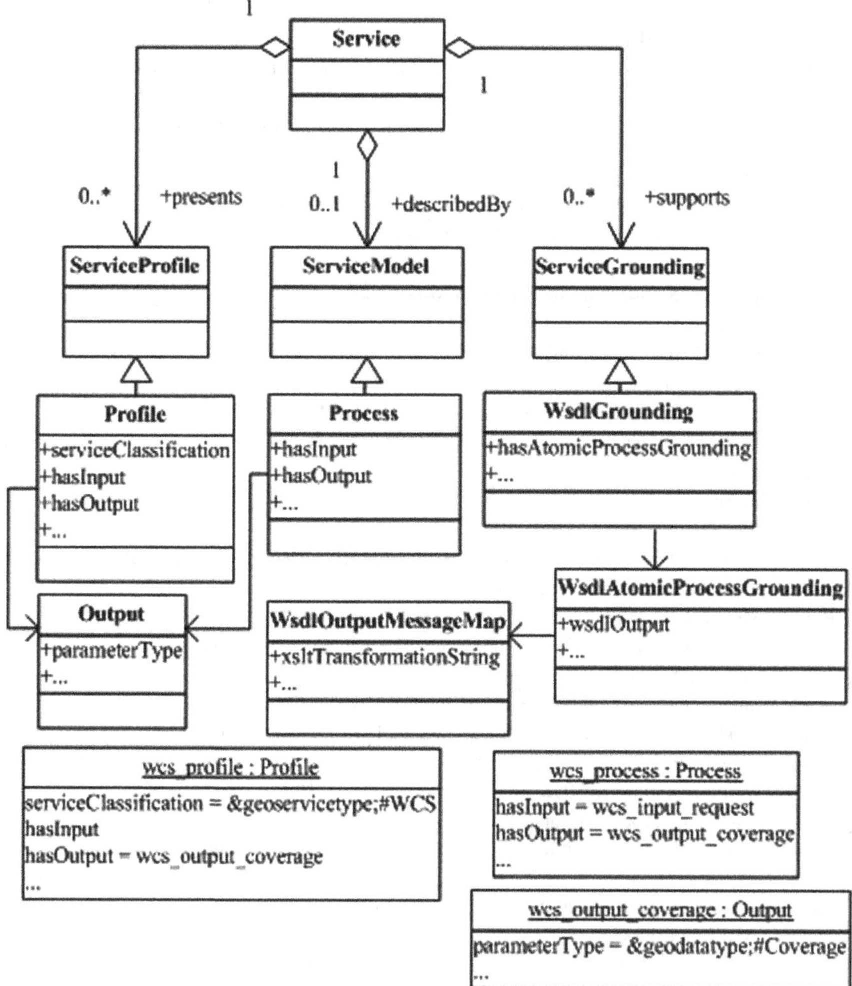

Fig. 5.4 OWL-S structure in the UML

geographic services. A large number of geospatial software tools, most of which do not have standard interface protocols, are available either as freeware or commercial products. These tools can be developed into Web services with non-OGC-compliant interfaces. Under these circumstances, OWL-S descriptions for these services need to be developed individually, based either on a specific service instance (e.g. slope service) or on a small aggregation of service instances from a certain software package (e.g. GRASS[3]). Hence the message mappings in the

[3] http://grass.itc.it/.

Table 5.1 A snippet of WSDL and OWL-S for the slope computation service

```
<!–snippet of Slope WSDL –>
<message name="DEM2SlopeRequest"><part name="sourceURL" type="xsd:anyURI"/
   >…</message>
<message name="DEM2SlopeResponse"><part name="DEM2SlopeReturnURL"
   type="xsd:anyURI"/>…</message>
<portType name="SlopeCal">…
<operation name="DEM2Slope"><input message="DEM2SlopeRequest"/>
<output message="DEM2SlopeResponse"/></operation></portType>

<!–snippet of OWL-S descriptions for Slope service –>
<!–Service description –>
<service:Service rdf:ID="slope_serice_01">
<service:describedBy rdf:resource="#slope_process_01"/>
<service:presents rdf:resource="#slope_profile_01"/>
<service:supports rdf:resource="#slope_wsdlgrounding_01"/>
</service:Service>
<!–Profile description –>
<profile:Profile rdf:ID=" slope_profile_01">
<profile:serviceClassification rdf:datatype="&xsd;#anyURI">&geoservicetype;#Slope
</profile:serviceClassification>…</profile:Profile>
<!–Process Model description –>
<process:AtomicProcess rdf:ID="slope_process_01">…</process:AtomicProcess>
<process:Input rdf:ID="slope_input_dem">
<process:parameterType rdf:datatype="&xsd;#anyURI">
&geodatatype;#Terrain_Elevation</process:parameterType></process:Input>
<!–Grounding description –>
<grounding:WsdlGrounding rdf:ID="slope_wsdlgrounding_01">
<grounding:hasAtomicProcessGrounding rdf:resource="#
   slope_wsdlatomicprocessgrounding_01"/></grounding:WsdlGrounding>

<!–Snippet of service grounding –>
<grounding:wsdlInputMessage
   rdf:datatype="&xsd;#anyURI">&slope_wsdl;#DEM2SlopeRequest
   </grounding:wsdlInputMessage>
<grounding:wsdlInput>
<grounding:WsdlInputMessageMap rdf:ID="slope_wsdlinputmessagemap_dataurl">
<grounding:owlsParameter rdf:resource="slope_input_dem"/>
<grounding:wsdlMessagePart rdf:datatype="&xsd;#anyURI">&slope_wsdl;#sourceURL
   </grounding:wsdlMessagePart>
<grounding:xsltTransformationString><![CDATA]
<xsl:stylesheet version="1.0" xmlns:xsl="http://www.w3.org/1999/XSL/Transform"
   xmlns:rdf="http://www.w3.org/1999/02/22-rdf-syntax-ns#" xmlns:iso19115="http://
   loki.cae.drexel.edu/~wbs/ontology/2004/09/iso-19115#"
```

(continued)

Table 5.1 (continued)

```
xmlns:mediator="http://www.laits.gmu.edu/geo/ontology/domain/v3/mediator_v3.owl#"
   xmlns:gedatatype="http://www.laits.gmu.edu/geo/ontology/domain/GeoDataType.owl#"
   xmlns="http://slope.laits.gmu.edu">
<xsl:template match="//geodatatype:Terrain_Elevation"><xsl:value-of
   select="mediator:hasMD_Metadata/iso19115:MD_Metadata/iso19115:distributionInfo/
   iso19115:MD_Distribution/iso19115:transferOptions/iso19115:MD_DigitalTransferOptions/
   iso19115:onLine/iso19115:CI_OnlineResource/iso19115:linkage"/>
</xsl:template></xsl:stylesheet>]]></grounding:xsltTransformationString>
</grounding:WsdlInputMessageMap><grounding:wsdlInput>
```

service grounding for non-OGC-compliant services need to be described case by case.

Until now execution semantics are not mentioned, while geospatial DataType and ServiceType ontologies can be used to address data semantics and functional semantics of geospatial services. When a thematic concept match (TBOX reasoning) based on the geospatial DataType ontology and ServiceType ontology is available, a geospatial service might still have multiple metadata constraint requirements such as file format, data projection on the input data. The execution semantics of a geospatial service can be specified using the metadata statement in the preconditions and effects. For example, the preconditions for a slope computation service in Fig. 5.3 specify that the input terrain elevation data should be in the GeoTIFF data format with the EPSG:4326 geographic coordinate reference system. This book proposes to define these metadata constraints in the OWL-S preconditions through the metadata structure. Before execution, a precondition check is required for the available data instances. Thus, precondition checking is in fact an ABOX reasoning problem. Tables 5.3 and 5.4 show the precondition definitions using SWRL and SPARQL respectively.

Introducing OWL-S into AI planning can be interpreted as: The state of the world is represented in the OWL knowledge base. And the OWL reasoner can be used to reason about the state of the world. Precondition checking is equivalent to querying the knowledge base to check the existence of some facts, e.g. the fact in Table 5.3 that the input precipitation data has file format HDFEOS. And applying effects is equivalent to adding and deleting facts from the knowledge base (Sirin et al. 2004). If the metadata constraints are the world states, then a service calculating the terrain slope from DEM data may require the HDFEOS data format as a precondition for DEM data.

5.2.3 Geospatial Semantics in Brokers

The broker contains information about information (meta-information) available over the Internet or in the holdings of digital libraries but not the information itself.

Table 5.2 Some examples on service grounding

	Grounding information
Type1 (values got from former service's output, e.g. WCS OWL-S output MessageMap)	<grounding:WsdlOutputMessageMap rdf:ID="wcs_wsdloutputmessagemap_coverage"> <grounding:owlsParameter rdf:resource="&wcs_profile;#wcs_output_ coverage"/> <grounding:xsltTransformationString><![CDATA] <xsl:stylesheet version="1.0" xmlns:xlink="http:// www.w3.org/ 1999/xlink" xmlns:wcs="http://www.opengis.net/wcs" xmlns:xsl= "http://www.w3.org/1999/XSL/Transform"> <xsl:template match="//wcs:Coverage/wcs:CoverageRegion/ wcs:CoverageData"> <xsl:variable name="X1" select="@xlink:href"/> <xsl:variable name="X2" select="wcs:Format"/> <rdf:RDF xmlns:rdf="http://www.w3.org/1999/02/22-rdf- syntax-ns#" xmlns:mediator="http://www.laits.gmu.edu/ geo/ontology/ domain/v3/mediator_v3.owl#" xmlns:iso19115=" http://loki.cae.drexel.edu/ ~ wbs/ontology/ 2004/09/ iso-19115#" xmlns:geodatatype="http:// www.laits.gmu.edu/geo/ ontology/domain/GeoDataType.owl#"> <geodatatype:Coverage> <mediator:hasMD_Metadata><iso19115:MD_Metadata> <iso19115:distributionInfo> <iso19115:MD_Distribution> <iso19115:transferOptions> <iso19115:MD_DigitalTransferOptions><iso19115:onLine> <iso19115:CI_OnlineResource><iso19115:linkage> <xsl:value-of select="$X1"/> </iso19115:linkage></iso19115:CI_OnlineResource> </iso19115:onLine></ iso19115:MD_DigitalTransferOptions> </iso19115:transferOptions> <iso19115:distributionFormat> <iso19115:MD_Format><iso19115:name_MD_Format> <xsl:value-of select="$X2"/> </iso19115:name_MD_Format></iso19115:MD_Format> </iso19115:distributionFormat> </iso19115:MD_Distribution> </iso19115:distributionInfo></iso19115:MD_Metadata></ mediator:hasMD_Metadata> </geodatatype:Coverage></rdf:RDF></xsl:template></ xsl:stylesheet>]]></grounding:xsltTransformationString> </grounding:WsdlOutputMessageMap>

(continued)

Table 5.2 (continued)

Type2 (values obtained from other ways such as hardcoded, e.g. one of WICS OWL-S input MessageMap)	<grounding:wsdlInput> <grounding:WsdlInputMessageMap rdf:ID="wics_mindis_train_wsdlinputmessagemap1"> <grounding:owlsParameter rdf:resource="#wics_mindis_train_input_service"/> <grounding:wsdlMessagePart rdf:datatype="&xsd;#anyURI">&wics_wsdl;#service </grounding:wsdlMessagePart> <grounding:xsltTransformationString><![CDATA] <xsl:stylesheet version="1.0" xmlns:xsl="http://www.w3.org/1999/XSL/Transform" xmlns:rdf="http://www.w3.org/1999/02/22-rdf-syntax-ns#" xmlns:geodatatype="http://www.laits.gmu.edu/geo/ontology/domain/GeoDataType.owl#" xmlns="http://www.opengis.net/wics"> <xsl:template match="/"> <xsl:text>WICS</xsl:text> </xsl:template> </xsl:stylesheet>]]></grounding:xsltTransformationString> </grounding:WsdlInputMessageMap> </grounding:wsdlInput>

The broker plays an important role in helping requestors to find the right services. Geospatial Web services are cataloged in a registry/broker with their properties and capabilities.

Currently, there are two prominent general models for registry services: the ebRIM and the UDDI model. For the geospatial community, ebRIM is more general and extensible because it provides comprehensive facilities, based on the ISO 11179 set of standards, to manage metadata. OGC has developed and recommended an ebRIM profile for CSW. This profile introduces an ebRIM-based catalogue information model for publication and discovery of geospatial information. The metadata for both geospatial Web services and geospatial data are registered in a CSW server.

While geospatial catalogue services greatly facilitate the discovery of data and services, the current discovery process is based on a static keyword match. The lack of explicit semantics inhibits the dynamic selection of those data, services, and geoprocessing workflows needed for processing geospatial information and discovering knowledge in a data-rich distributed environment.

The ebRIM model is a general information model. It provides standard mechanisms to define and associate semantic information with registered information resources. Such mechanisms include using a cohesive set of extensibility points such as new kinds of associations, classifications, and additional slots (more details in Chap. 6). On the other hand, the Semantic Web is a separate effort. Semantics for geospatial data, services and geoprocessing service chains are

Table 5.3 An example of file format requirement represented in the OWL-S precondition using SWRL

```
<expr:SWRL-Condition rdf:ID=" supportedFileFormat">
<rdfs:label>(input_precipitation_amount &mediator;#hasMD_Metadata md_metadata)
   (md_metadata &iso19115;#distributionInfo md_distribution) (md_distribution
   &iso19115;#distributionFormat file_format)</rdfs:label>
<expr:expressionLanguage rdf:resource="&expr;#SWRL"/>
<expr:expressionBody rdf:parseType="Literal">
<swrl:AtomList><rdf:first><swrl:ClassAtom><swrl:classPredicate
   rdf:resource="&fileformat;#HDFEOS"/>
<swrl:argument1><swrl:Variable rdf:ID="file_format"/>
</swrl:argument1></swrl:ClassAtom></rdf:first>
<rdf:rest><swrl:AtomList><rdf:first><swrl:ClassAtom>
<swrl:classPredicate rdf:resource="&iso19115;#MD_Metadata"/>
<swrl:argument1><swrl:Variable rdf:ID="md_metadata"/>
</swrl:argument1></swrl:ClassAtom></rdf:first>
<rdf:rest><swrl:AtomList><rdf:first><swrl:ClassAtom>
<swrl:classPredicate rdf:resource="&iso19115;#MD_Distribution"/>
<swrl:argument1><swrl:Variable rdf:ID="md_distribution"/>
</swrl:argument1></swrl:ClassAtom></rdf:first><rdf:rest>
<swrl:AtomList><rdf:first><swrl:IndividualPropertyAtom>
<swrl:propertyPredicate rdf:resource="&mediator;#hasMD_Metadata"/>
<swrl:argument1 rdf:resource="# input_precipitation_amount "/>
<swrl:argument2 rdf:resource="#md_metadata"/>
</swrl:IndividualPropertyAtom></rdf:first><rdf:rest>
<swrl:AtomList><rdf:first><swrl:IndividualPropertyAtom>
<swrl:propertyPredicate rdf:resource="&iso19115;#distributionInfo"/>
<swrl:argument1 rdf:resource="#md_metadata"/>
<swrl:argument2 rdf:resource="#md_distribution"/>
</swrl:IndividualPropertyAtom></rdf:first><rdf:rest>
<swrl:AtomList><rdf:first><swrl:IndividualPropertyAtom> <swrl:propertyPredicate
   rdf:resource="&iso19115;#distributionFormat"/>
<swrl:argument1 rdf:resource="#md_distribution"/>
<swrl:argument2 rdf:resource="#file_format"/>
</swrl:IndividualPropertyAtom></rdf:first>
<rdf:rest rdf:resource="&rdf;#nil"/></swrl:AtomList>
</rdf:rest></swrl:AtomList></rdf:rest></swrl:AtomList>
</rdf:rest></swrl:AtomList></rdf:rest></swrl:AtomList>
</rdf:rest></swrl:AtomList></expr:expressionBody>
</expr:SWRL-Condition>
```

represented using OWL/OWL-S. An important initiative for semantics-enhanced discovery of information resources based on ebRIM is to incorporate these explicitly defined semantics in OWL/OWL-S into ebRIM using these extensibility points. Various constructs in OWL are mapped to different ebRIM elements. Several efforts have already addressed this issue, although focusing only on the general information domain (Dogac 2006; Dogac et al. 2005; Liu et al. 2005;

Table 5.4 An example of file format requirement represented in the OWL-S precondition using SPARQL

```
<expr:SPARQL-Condition rdf:ID="supportedFileFormat">
<expr:expressionLanguage rdf:resource="&expr;#SPARQL"/>
<expr:expressionBody rdf:parseType="Literal">
<sparqlQuery xmlns="http://www.w3.org/2002/ws/sawsdl/spec">
PREFIX iso19115: &lt;http://loki.cae.drexel.edu/~wbs/ontology/2004/09/iso-19115#&gt;
PREFIX mediator: &lt;http://www.laits.gmu.edu/geo/ontology/domain/v3/
    mediator_v3.owl#&gt;
PREFIX fileformat: &lt;http://www.laits.gmu.edu/geo/ontology/domain/v2/fileformat.owl#&gt;
PREFIX rdf: &lt;http://www.w3.org/1999/02/22-rdf-syntax-ns#&gt;
SELECT
WHERE {
?coverage mediator:hasMD_Metadata ?md_metadata.
?md_metadata rdf:type iso19115: MD_Metadata .
?md_metadata iso19115:distributionInfo ?md_disinfo.
? md_disinfo rdf:type iso19115: MD_Distribution .
?md_disinfo iso19115:distributionFormat ?file_format.
?file_format rdf:type fileformat:HDFEOS}
  </sparqlQuery>
</expr:expressionBody>
<expr:variableBinding>
<expr:VariableBinding>
<expr:theVariable>coverage</expr:theVariable>
<expr:theObject rdf:resource="#wildfireprediction_input_maxt"/>
</expr:VariableBinding>
</expr:variableBinding>
</expr:SPARQL-Condition>
```

Bechini et al. 2008). Chapter 6 will discuss how to make extensions to the ebRIM information model for geospatial catalogue services.

5.2.4 Geospatial Semantics in Requestors

A requestor represents the consumer or user of Web services who needs geospatial information. A user may request a service to generate a data product, or may request a data product without knowing the specific service(s) needed to generate the product. The latter case is convenient to the general geospatial users. In the design, a user request is expressed by a concept in the geospatial DataType ontology, which represents the content or theme of the requested product. In addition to the geospatial DataType, a geospatial query is often associated with other conditions, especially temporal and spatial constraints. Therefore, a complete query consists of at least three major elements, a geospatial DataType concept representing the content of the query, a temporal domain, and a spatial domain. Table 5.5 is an example of such a request in XML generated for the use case 1 in

Table 5.5 A sample request

```
<TimeRange>
<Start>2005-01-10T00:00:00Z</Start>
<End>2005-01-10T23:59:59Z</End>
</TimeRange>
<PlaceBoundingBox crs="EPSG:4326">
<WestBoundingLongitude>-122.262908</WestBoundingLongitude>
<SouthBoundingLatitude>37.597494</SouthBoundingLatitude>
<EastBoundingLongitude>−122.005009</EastBoundingLongitude>
<NorthBoundingLatitude>37.875999</NorthBoundingLatitude>
</PlaceBoundingBox>
<Ontology>http://www.laits.gmu.edu/geo/ontology/domain/
  GeoDataType.owl#Landslide_Susceptibility</Ontology>
```

Chap. 1. This XML specifies the temporal/spatial ranges during/among which the information is requested. The Ontology element of the XML specifies the type of information (i.e., a geospatial DataType). Through the transformations such as XSLT, it can be transformed into an ontology entity in Table 5.6.

5.2.5 Geospatial Semantics in Service Chain

The XML-based service composition languages such as BPEL rely on the XML and XML schema descriptions of individual Web services for constructing service chains. Certain schema-match mechanisms are required for enabling the chaining of Web services with heterogeneous interfaces and messages. For example, in order to chain a WCS service that provides DEM data and a Slope service, a non-OGC-compliant service defined by the service provider that generates slope data from DEM (Fig. 5.5), we need first to extract the data URL and data format from the "Coverage" message structure defined in the OGC WCS schema, and then transfer them to the "souceURL" and "sourceFormat" parts of the DEM2Slope-Request message in the Slope service. Through the input/output XSLT transformation defined in the service grounding of OWL-S, this value-transfer process can be performed automatically at run time.

When two services are chained, there must be a mapping between the message schemas of the services. One approach is to define direct schema mapping among all available services. In a Web environment where n services are available, the maximum possible number of such mappings is $C(n,2)$. For standards-compliant services, the mappings can be defined at the service type level rather than at the service instance level, which reduces the number $C(n,2)$ to $C(m,2)$, where m, representing the number of service types to which the n service instances belong, is usually much smaller than n. For services not compliant with standards and thus without standard interface schemas, the number of direct schema mappings between each pairs of chainable services can be much larger. With the introduction of geospatial ontology, the mapping number can be reduced from $C(n,2)$ to

Table 5.6 An example of desired data product


```
<geodatatype:Landslide_Susceptibility
xmlns:geodatatype="http://www.laits.gmu.edu/geo/ontology/domain/GeoDataType.owl#"
xmlns:gml="http://www.opengis.net/gml"
xmlns:mediator="http://www.laits.gmu.edu/geo/ontology/domain/v3/mediator_v3.owl#"
xmlns:gml-ont="http://loki.cae.drexel.edu/~wbs/ontology/2004/09/ogc-gml#"
xmlns:gmlpacket="http://www.opengis.net/examples/packet"
xmlns:iso19115="http://loki.cae.drexel.edu/~wbs/ontology/2004/09/iso-19115#"
xmlns:iso19112="http://loki.cae.drexel.edu/~wbs/ontology/2004/09/iso-19112#"
xmlns:iso19103="http://loki.cae.drexel.edu/~wbs/ontology/2004/09/iso-19103#"
xmlns:iso19107="http://loki.cae.drexel.edu/~wbs/ontology/2004/09/iso-19107#"
xmlns:iso19108="http://loki.cae.drexel.edu/~wbs/ontology/2004/09/iso-19108#"
xmlns:rdf="http://www.w3.org/1999/02/22-rdf-syntax-ns#">
<mediator:hasMD_Metadata>
<iso19115:MD_Metadata>
<iso19115:identificationInfo>
<iso19115:MD_DataIdentification><iso19115:dataExtent><iso19115:EX_Extent>
<iso19115:geographicElement><iso19115:EX_GeographicBoundingBox>
<iso19115:westBoundLongitude>
<iso19103:Angle><iso19103:value>-122.262908</iso19103:value></iso19103:Angle>
</iso19115:westBoundLongitude>
<iso19115:eastBoundLongitude>
<iso19103:Angle><iso19103:value>-122.005009</iso19103:value></iso19103:Angle>
</iso19115:eastBoundLongitude>
<iso19115:southBoundLatitude>
<iso19103:Angle><iso19103:value>37.597494</iso19103:value></iso19103:Angle>
</iso19115:southBoundLatitude>
<iso19115:northBoundLatitude>
<iso19103:Angle><iso19103:value>37.875999</iso19103:value></iso19103:Angle>
</iso19115:northBoundLatitude>
</iso19115:EX_GeographicBoundingBox></iso19115:geographicElement>
<iso19115:temporalElement>
<iso19115:EX_TemporalExtent><iso19115:exTemp><iso19108:TM_Period>
<iso19108:beginning>
<iso19108:TM_Instant><iso19108:position><iso19108:TM_Position_DateTime8601>
<iso19108:dateTime8601>2005-01-10T00:00:00Z</iso19108:dateTime8601>
</iso19108:TM_Position_DateTime8601></iso19108:position></iso19108:TM_Instant>
</iso19108:beginning>
<iso19108:ending>
<iso19108:TM_Instant><iso19108:position><iso19108:TM_Position_DateTime8601>
<iso19108:dateTime8601>2005-01-10T23:59:59Z</iso19108:dateTime8601>
</iso19108:TM_Position_DateTime8601></iso19108:position></iso19108:TM_Instant>
</iso19108:ending>
</iso19108:TM_Period></iso19115:exTemp></iso19115:EX_TemporalExtent>
</iso19115:temporalElement>
</iso19115:EX_Extent></iso19115:dataExtent></iso19115:MD_DataIdentification>
</iso19115:identificationInfo>
</iso19115:MD_Metadata>
</mediator:hasMD_Metadata>
</geodatatype:Landslide_Susceptibility>
```

Table 5.7 XSLT example from GML to Drexel GML ontology

```
<!– from GML polygon to Polygon ontology –>
<xsl:template match=“gml:Polygon”>
<xsl:element name=“gml-ont:Polygon” xmlns:gml-ont=“http://loki.cae.drexel.edu/~wbs/
   ontology/2004/09/ogc-gml#”>
<xsl:apply-templates select=“gml:outerBoundaryIs”/>
<xsl:apply-templates select=“gml:innerBoundaryIs”/></xsl:element>
</xsl:template>

<xsl:template match=“gml:outerBoundaryIs”>
<xsl:element name=“gml-ont:exterior_Ring” xmlns:gml-ont=“http://loki.cae.drexel.edu/~wbs/
   ontology/2004/09/ogc-gml#”>
<xsl:apply-templates select=“gml:LinearRing”/> </xsl:element>
</xsl:template>

<xsl:template match=“gml:innerBoundaryIs”>
<xsl:element name=“gml-ont:interior_Ring” xmlns:gml-ont=“http://loki.cae.drexel.edu/~wbs/
   ontology/2004/09/ogc-gml#”>
<xsl:apply-templates select=“gml:LinearRing”/></xsl:element>
</xsl:template>

<xsl:template match=“gml:LinearRing”>
<xsl:element name=“gml-ont:LinearRing” xmlns:gml-ont=“http://loki.cae.drexel.edu/~wbs/
   ontology/2004/09/ogc-gml#”>
<gml-ont:positions><xsl:call-template name=“add-points”>
<xsl:with-param name=“nodeset” select=“gml:coord”/>
</xsl:call-template></gml-ont:positions></xsl:element>
</xsl:template>

<!– from GML lineString to LineString ontology –>
<xsl:template match=“gml:LineString”>
<xsl:element name=“gml-ont:LineString” xmlns:gml-ont=“http://loki.cae.drexel.edu/~wbs/
   ontology/2004/09/ogc-gml#”>
<xsl:call-template name=“add-points”><xsl:with-param name=“nodeset” select=“gml:coord”/
   >
</xsl:call-template></xsl:element>
</xsl:template>

<!– from gml coord to RDF:List–>
<xsl:template name=“add-points”>
<xsl:param name=“nodeset”/><rdf:List><xsl:choose><xsl:when test=“count($nodeset) &gt;
   1”>
<rdf:first><gml-ont:Position_coord><gml-ont:coord><gml-ont:Coord><gml-ont:X>
<xsl:value-of select=“$nodeset[1]/gml:X”/></gml-ont:X><gml-ont:Y>
<xsl:value-of select=“$nodeset[1]/gml:Y”/></gml-ont:Y></gml-ont:Coord></gml-ont:coord>
</gml-ont:Position_coord></rdf:first><rdf:rest>
```

(continued)

Table 5.7 (continued)

```
<xsl:call-template name="add-points"><xsl:with-param name="nodeset"
    select="$nodeset[position()>1]"/>
</xsl:call-template></rdf:rest></xsl:when><xsl:otherwise><rdf:first>
<gml-ont:Position_coord><gml-ont:coord><gml-ont:Coord><gml-ont:X><xsl:value-of
    select="$nodeset[1]/gml:X"/></gml-ont:X><gml-ont:Y><xsl:value-of
    select="$nodeset[1]/gml:Y"/></gml-ont:Y></gml-ont:Coord>
</gml-ont:coord></gml-ont:Position_coord></rdf:first>
<rdf:rest rdf:resource="http://www.w3.org/1999/02/22-rdf-syntax-ns#nil"/>
</xsl:otherwise></xsl:choose></rdf:List>
</xsl:template>

<!– from GML point to Point ontology –>
<xsl:template match="gml:Point">
<xsl:element name="gml-ont:Point" xmlns:gml-ont="http://loki.cae.drexel.edu/~wbs/
    ontology/2004/09/ogc-gml#">

<gml-ont:position><gml-ont:DirectPositionChoice_coord>
<gml-ont:coord><gml-ont:Coord><gml-ont:X><xsl:value-of select="gml:coord/gml:X"/>
</gml-ont:X><gml-ont:Y><xsl:value-of select="gml:coord/gml:Y"/>
</gml-ont:Y></gml-ont:Coord></gml-ont:coord></gml-ont:DirectPositionChoice_coord>
</gml-ont:position></xsl:element>
</xsl:template>
```

n because messaging mappings are indirectly embodied in the mapping of the services message schema structure to a mediated RDF structure.

The mediated RDF structure is defined by enriching the "DataType" ontology with the ISO 19115 ontology and GML ontology developed by Drexel University (Drexel 2004). Addition properties are defined, including "hasMD_Metadata" and "hasGML", so that each "DataType" has standards-based semantic metadata and formalized geometry concepts. "GeoDataType" serves as the top level concept of "DatatType" ontology. Tables 5.7 and 5.8 show examples of XSLT between GML and the GML ontology. These transformations can be imported in the service grounding of any OWL-S descriptions for geospatial Web services with GML parameters.

Aalst (2003) compared several common service composition languages from the aspect of control flow. Twenty flow control constructs, such as sequence, parallel split, and choice, were identified as the considerations most often required when designing a service composition language. OWL-S provides a "Composite Process" ontology with control constructs for these pattern definitions. A process can be either atomic or composite. Both atomic and composite processes can be advertised through service profile ontology by their functionalities, inputs, outputs, preconditions, and effects. Atomic process ontology in OWL-S describes the behavior of an atomic service, while a composite process is a collection of subprocesses or atomic processes with control and data flow relationships. Therefore, the semantics for a geospatial service chain can be represented using composite process ontology.

Table 5.8 XSLT example from Drexel GML ontology to GML

```
<!– from Polygon ontology to GML polygon –>
<xsl:template match=“gml-ont:Polygon”>
<xsl:element name=“gml:Polygon” xmlns:gml=‘http://www.opengis.net/gml’>
<gml:outerBoundaryIs><xsl:apply-templates select=“gml-ont:exterior_Ring/gml-
   ont:LinearRing”/>
</gml:outerBoundaryIs>
<gml:innerBoundaryIs><xsl:apply-templates select=“gml-ont:interior_Ring/gml-
   ont:LinearRing”/>
</gml:innerBoundaryIs></xsl:element>
</xsl:template>

<xsl:template match=“gml-ont:LinearRing”>
<xsl:element name=“gml:LinearRing” xmlns:gml=http://www.opengis.net/gml’>
<xsl:call-template name=“add-points”><xsl:with-param name=“rdflist” select=“gml-
   ont:positions/rdf:List”/>
</xsl:call-template></xsl:element>
</xsl:template>

<!– from LineString ontology to GML lineString –>
<xsl:template match=“gml-ont:LineString”>
<xsl:element name=“gml:LineString” xmlns:gml=‘http://www.opengis.net/gml’>
<xsl:call-template name=“add-points”><xsl:with-param name=“rdflist” select=“gml-
   ont:positions/rdf:List”/>
</xsl:call-template></xsl:element>
</xsl:template>

<!– from RDF:List to gml coord –>
<xsl:template name=“add-points”>
<xsl:param name=“rdflist”/><xsl:choose>
<xsl:when test=“$rdflist/rdf:rest/rdf:List”><gml:coord><gml:X>
<xsl:value-of select=“$rdflist/rdf:first/gml-ont:Position_coord/gml-ont:coord/gml-ont:Coord/
   gml-ont:X”/>
</gml:X><gml:Y>
<xsl:value-of select=“$rdflist/rdf:first/gml-ont:Position_coord/gml-ont:coord/gml-ont:Coord/
   gml-ont:Y”/>
</gml:Y></gml:coord><xsl:call-template name=“add-points”>
<xsl:with-param name=“rdflist” select=“$rdflist/rdf:rest/rdf:List”/>
</xsl:call-template></xsl:when>
<xsl:otherwise><gml:coord><gml:X>
<xsl:value-of select=“$rdflist/rdf:first/gml-ont:Position_coord/gml-ont:coord/gml-ont:Coord/
   gml-ont:X”/>
</gml:X><gml:Y>
<xsl:value-of select=“$rdflist/rdf:first/gml-ont:Position_coord/gml-ont:coord/gml-ont:Coord/
   gml-ont:Y”/>
</gml:Y></gml:coord></xsl:otherwise></xsl:choose>
```

(continued)

Table 5.8 (continued)

```
</xsl:template>

<!– from Point ontology to GML point –>
<xsl:template match="gml-ont:Point">
<xsl:element name="gml:Point" xmlns:gml='http://www.opengis.net/gml'>
<gml:coord><gml:X>
<xsl:value-of select="gml-ont:position/gml-ont:DirectPositionChoice_coord/gml-ont:coord/gml-
   ont:Coord/gml-ont:X"/>
</gml:X><gml:Y>
<xsl:value-of select="gml-ont:position/gml-ont:DirectPositionChoice_coord/gml-ont:coord/gml-
   ont:Coord/gml-ont:Y"/>
</gml:Y></gml:coord></xsl:element>
</xsl:template>
```

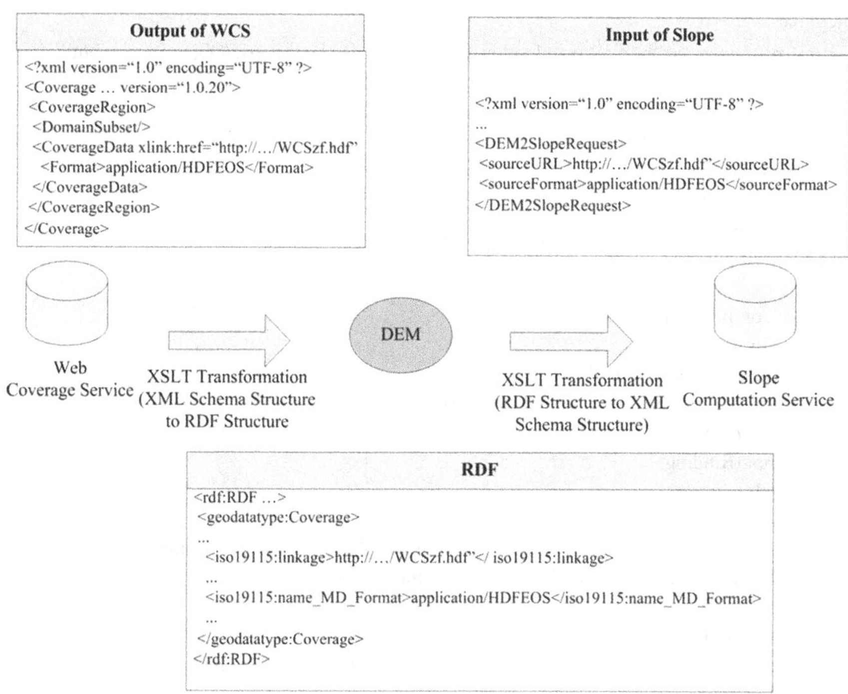

Fig. 5.5 Data flow in the service chain

Linking geospatial DataTypes, ServiceTypes, and workflow ontologies together, semantics for geospatial service chains are represented (Fig. 5.3). Table 5.9 illustrates semantic descriptions for the landslide susceptibility case using workflow ontologies in OWL-S. The control flow is represented by the control constructs such as Sequence and Split-Join. The data flow is specified through input/output bindings using a class such as ValueOf to state that the input to one subprocess should be the

Table 5.9 A snippet of OWL-S for a geoprocessing workflow

```
<!– snippet of a composite process –>
<!– Control Flow –>
<process:CompositeProcess …>
<process:composedOf>
<process:Sequence>
<process:components>
<process:ControlConstructList>
<list:first>
<process:Split-Join>
<process:components>
<process:ControlConstructBag > …
<list:rest>
<process:ControlConstructList>
<list:first>
<process:Perform rdf:nodeID="A0"/>
</list:first>
<list:rest rdf:resource="http://www.daml.org/services/owl-s/1.1/generic/ObjectList.owl#nil"/>
</process:ControlConstructList>
</list:rest>
</process:ControlConstructList>
</process:components>
</process:Sequence>
</process:composedOf>
<process:hasInput …/>
...
</process:CompositeProcess>
<!– Data Flow –>
<process:Perform rdf:nodeID="A0">
<process:process rdf:resource="&landslide_sus_4i;#landslide_sus_4i_process_01"/>
<process:hasDataFrom>
<process:InputBinding>
<process:valueSource>
<process:ValueOf>
<process:fromProcess><process:Perform rdf:nodeID="A7"/></process:fromProcess>
<process:theVar rdf:resource="&slope;#slope_output_slope"/>
</process:ValueOf>
</process:valueSource>
<process:toParam rdf:resource="&landslide_sus_4i;#landslide_sus_4i_input_slope"/>
</process:InputBinding>
</process:hasDataFrom>
<process:hasDataFrom>…
</process:Perform>
<process:Perform rdf:nodeID="…">…</process:Perform>
...
```

output of the previous one. For example, as shown in Table 5.9, the output (slope_output_slope) of the slope computation process is linked to the input (landslide_sus_4i_input_slope) of the landslide susceptibility atomic process.

Composite processes are processes decomposable into other (non-composite or composite) processes. The decomposition can be specified by using control constructs. Since most control construct definitions originate from the service composition languages, it is possible for business processes defined in any of the service composition languages to map to the “Composite Process” ontology, thus achieving interoperability between service composition languages. Also, a composite process in OWL-S resulting from service composition can be converted to any of the service composition languages to enable execution in existing engines of these languages.

Chapter 6
Semantics-Enabled Geospatial Data and Services Discovery

6.1 An ebRIM Profile of Geospatial Catalogue Service

In the geospatial domain, a geospatial catalogue service provides a network-based meta-information repository and interface for advertising and discovering shared geospatial data and services. OGC technology is the widely used choice for the standards-based interoperability and sharing technology in the geospatial domain. The most widely used interface specification for geospatial catalogue services is the OGC CSW. It is an open industry consensus on a standard interface to online catalogs for geospatial data, services, and related resource information. Descriptive information (i.e., metadata) for geospatial information resources is structured and organized in catalogue services. The metadata can be queried and returned for evaluation, processing, and further binding or invocation of the cited resource. However, current standards mainly focus on syntactic interoperability and do not address semantic interoperability (ISO/TC211 2005). This work uses OGC standards to address the semantic interoperability of geospatial catalogue services.

Figure 6.1 shows the relations among the OGC catalogue services, CSW, and the ebRIM profile of CSW. The core elements in an OGC catalogue service are the information model, the query language, and the interface (Nebert et al. 2007). The *information model* describes the *information structures and semantics* of information resources. Therefore, the information model of catalogue services should address the content, syntax, and semantics of geospatial data, services, and geoprocessing service chains. The OGC catalogue specification is a general framework for catalogue service implementation. Application profiles can be derived from this base specification (Nebert et al. 2007). Interoperability among the different profiles requires the specification of a set of core metadata elements in the information model, in particular, the *core queryable properties* and *common returnable properties*. For example, the spatial extent is such a core metadata element. It is represented by a Bounding Box element in the core queryable properties and a coverage element (interpreted as the Bounding Box in the context of metadata for geospatial data and services) in the common returnable properties (Nebert et al. 2007). Queries based on these core queryable properties can be

P. Yue, *Semantic Web-based Intelligent Geospatial Web Services*,
SpringerBriefs in Computer Science, DOI: 10.1007/978-1-4614-6809-7_6,

executed by any catalogue service, while the common returnable properties permit the use of metadata from any catalogue service. The *query language* assists in discovery of information resources in the catalogue. Different implementations of query languages, such as the OGC Filter Specification or Catalogue Interoperability Protocol (CIP) and GEO profiles of Z39.50 Type-1 queries, should support a minimum set of data types and query operations, the so-called OGC_Common Catalogue Query Language, to allow interoperability. For example, the OGC_Common Catalogue Query Language defines spatial operators such as Intersects and Within that should be supported by all query language implementations to determine whether geometric arguments satisfy the claimed spatial relationship. The *interface* defines the functional behaviors of the catalogue service such as discovery and transactional operations. For example, it includes the get Capabilities operation, an operation supported by most OGC service specifications that allows clients to retrieve service metadata. Implementation of the interface in different distributed computing environments results in different protocol bindings, e.g., the CORBA protocol binding and HTTP protocol binding. CSW is a specification focusing on operations in the Web environment. It follows the HTTP protocol binding and can support XML encoding of the OGC Filter query language. The ebRIM standard has been defined by OASIS and selected by OGC as the information model for specifying how catalogue content is structured and interrelated. Therefore, OGC proposes and recommends an ebRIM profile of CSW to join the CSW interfaces to ebRIM ((Martell 2008).

The ebRIM model specifies the metadata for information resources by using a set of classes and relationships among these classes. The UML style graph of Fig. 6.2 shows relationships of the metadata classes defined by the model (OASIS 2005). The core metadata class is the RegistryObject. Most other metadata classes in the information model are derived from this class. An instance of RegistryObject may have a set of zero or more Slot instances that serves as extensible attributes for this RegistryObject instance. An Association instance represents an association between a source RegistryObject and a target RegistryObject. Each

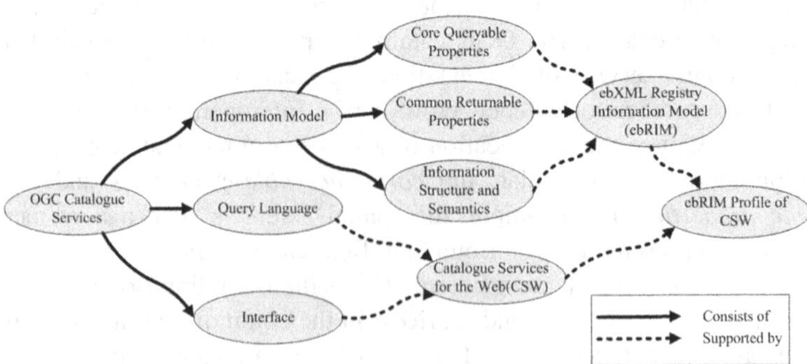

Fig. 6.1 OGC catalogue services and the ebRIM profile of CSW

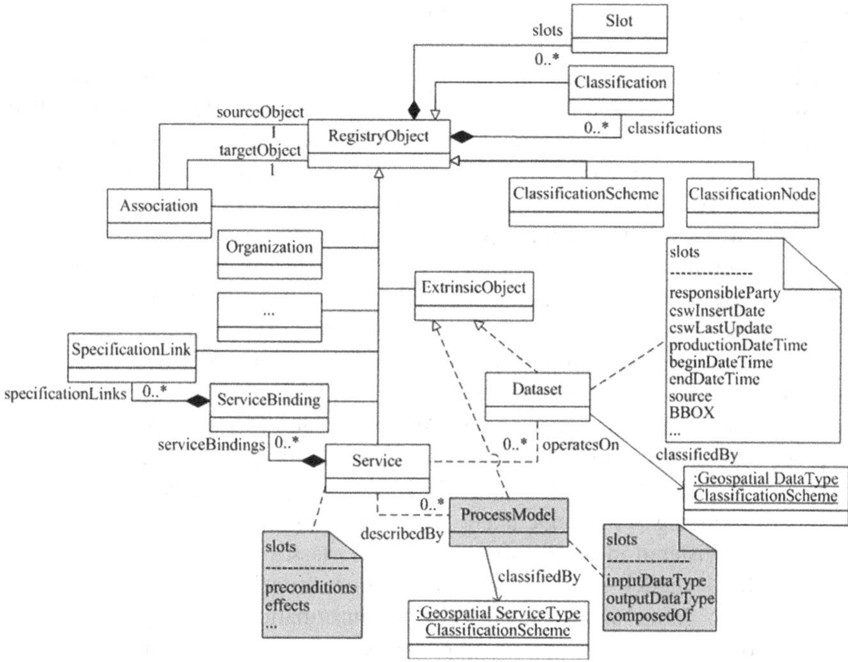

Fig. 6.2 Extension to the ebRIM information model

association has an association Type attribute that identifies the type of that association.

The classification mechanism is a significant feature in the ebRIM information model. A Classification instance classifies a RegistryObject instance by referring to a node defined within a ClassificationScheme instance. A ClassificationScheme instance in the ebRIM model defines a tree structure made up of nodes that can be used to describe a taxonomy. The structure of a classification scheme may be defined internally to or externally of the registry, resulting in a distinction between internal and external classification schemes. The nodes in an internal classification scheme are instances of ClassificationNode. In an external classification scheme, the structure and values of the taxonomy elements are not known to the Registry. Classifications could be internal or external, depending on whether the classification scheme used is internal or external. The attributes in the Classification class allow for representation of both internal and external classifications (OASIS 2005). An internal classification refers to a ClassificationNode in the internal ClassificationScheme, while an external classification refers to the node indirectly by specifying a representation of the node value unique within the external classificationScheme.

The ebRIM model is a general standard model that can be adapted to meet specific requirements in the geospatial domain. The CSW-ebRIM profile (Martell 2008) has provided guidance for registration of geospatial metadata by taking advantage of extensibility points offered by ebRIM. These extensibility points include new types of ExtrinsicObject, new kinds of associations, classifications, and additional slots. The dashed lines in Fig. 6.2 show extensions using these extensibility points. ExtrinsicObject provide metadata that describes submitted content whose type is not intrinsically known to the registry and therefore must be described by means of additional attributes. For example, metadata for geospatial data can be registered by creating a new type of ExtrinsicObject, i.e. Dataset (Fig. 6.2). New attributes such as spatial and temporal properties can be added to Dataset by defining additional slots. The ebRIM model has provided the Service class that supports the registration of service descriptions. A service chain as a whole can be conceived of as having a single-step execution that has inputs/outputs and performs a complex function. A WSDL can be also defined for a service chain. For example, BPEL is an industry-wide standard that can be used for syntactic specification of service chains. An executable BPEL process can provide the process description for a service chain using activities, partners, and messages exchanged between these partners. A BPEL process works as a Web service and has a corresponding WSDL document. Therefore, descriptions of a service chain can also be registered in CSW using the Service class.

The ebRIM model is a general information model. It provides standard mechanisms to define and associate semantic information with registered information resources. Such mechanisms include using a cohesive set of extensibility points such as new kinds of associations, classifications, and additional slots. On the other hand, the Semantic Web is a separate effort. Semantics for geospatial data, services, and geoprocessing service chains are represented using OWL/OWL-S. An important initiative for semantics-enhanced discovery of information resources based on ebRIM is to incorporate these explicitly defined semantics in OWL/OWL-S into ebRIM using these extensibility points. Various constructs in OWL are mapped to different ebRIM elements. Several efforts have already addressed this issue, although focusing only on the general information domain (Dogac 2006; Dogac et al. 2005; Liu et al. 2005). The next section will discuss related work, followed by solutions on how to make extensions to the ebRIM information model for geospatial catalogue services in Sect. 6.3.

6.2 Semantics-Enabled Service Registry: Current Solutions

The service registry plays an important role in helping requestors to find the right services. Web services are cataloged in a registry/broker with their properties and capabilities. As mentioned before, there are two prominent general models for registry services: ebRIM and UDDI. Currently, the search functionality for both of them is limited to the direct match of keywords from metadata without fully utilizing the semantic information implicitly embedded in the metadata, such as hierarchical relationships among metadata entities. Plus there is no search mechanism based on the capabilities of services in terms of services' operation/functionality, input/output and pre/post-conditions. Thus some efforts are trying to adding semantics information into UDDI or ebRIM to enable a semantics-enabled search.

6.2.1 Adding Semantics into UDDI

In the general information domain, much work has been conducted on adding semantics to UDDI. Paolucci et al. (2002a) introduce a mapping from the OWL-S to UDDI data model. UDDI describes three types of entities: (a) Business Entity, which records contact and owner information; (b) Business Service, which describes one or more specific services that a business provides; (c) Binding Template, which specifies the service access end point. In addition to these entities, UDDI provides a data structure called TModel that can specify the additional attributes of entities, thus allowing description of the specified ontological concepts. Each service can have one or more TModels that help describe its characteristics. Thus, service capabilities such as function or service input/output can be recorded in the corresponding TModels. Currently, most other efforts use the similar mapping and differ in the implementation of semantic search component.

There are three options are available now for the implementation of semantic search functionality.

(1) Option 1: The functionality is created outside of the registry, without any change to the registry interface (Paolucci et al. 2002a, b; Sivashanmugam et al. 2003; Srinivasan et al. 2004). An OWL-S Matching Engine is developed to handle the semantics-enabled search (Paolucci et al. 2002a).

Steps for the registration of service semantics: (a) Advertises services in the form of OWL-S; (b) Based on the mapping of OWL-S profile to the UDDI data model, constructs the UDDI service description using information in the OWL-S and registers it into the UDDI. (c) Gets the reference ID of the service from the result of registration with UDDI, combine it with the capabilities description of service advertisement, and store them into the AdvertisementDB (Advertisement Data Base) component of the OWL-S Matching Engine.

Steps of semantics-enabled service discovery: (a) Constructs the service request in the form of OWL-S. (b) OWL-S Matching Engine selects the advertisements from the AdvertisementDB and computes the level of match to the request based on the output-first and input-second semantic match (Paolucci et al. 2002b). (c) Gets the UDDI records based on the reference ID from the matching result and combines them with the advertisement from the matching result as the response.

Due to the possible huge amount of advertisements, the matching process will be extremely time-consuming. A pre-computation of match in the publishing phase of service is adopted where each ontological concept is indexed with the related services and their match level at the input or output. Since the matching information is pre-computed at the publishing phase, the query phase is reduced to simple lookups in the hierarchical data structure (Srinivasan et al. 2004).

Instead of mapping OWL-S to UDDI structures, Sivashanmugam et al. (2003) introduces the mapping from WSDL-S to UDDI structures, while the design of TModels in the UDDI is still similar. And they enhance the matching ability in the (Paolucci et al. 2002b) with the consideration of functionality of service operations. First, services are selected based on the ontological concepts of functionality, then they are pruned using the input and output match.

(2) Option 2: Semantic search functionality is embedded into the registry with some changes to the registry interface to support the semantically augmented query, for example, a RDF representation is embedded in the UDDI query. The UDDI API schema is extended with a property (RDF: Property) referring to the ontological concepts (Akkiraju et al. 2003). The service publishing steps are similar to the (Paolucci et al. 2002a), except that it does not maintain an AdvertisementDB. The service discovery steps are as follows: (a) Constructs the service request following the UDDI API schema (at this time it contains the semantic information according to the schema extension). (b) Gets the filtered set of services according to those filters of standard UDDI schema (the standard UDDI find method can be used). (c) Filtered set of services are sent to the semantic matching engine to enable the semantic match with the requested ontological concepts. The match is based on the input and output match. If no match is available, the semantic matching engine will compose services to meet the original request.

(3) Option 3: The functionality is wrapped as an individual external matching service registered in the registry. In this option, UDDI relays the matching task to the external matching services to enable the different types of matching such as OWL-S, WSDL, and UML (Colgrave et al. 2004). The registered service information includes the identification of its appropriate external matching information. The service discovery process includes three stages: (a) Detects the need for external matching from the request and takes it as a filter to retrieve the relevant external matching description of services. (b) Looks for available and compatible external matching services and invokes the appropriate external matching service by passing the requirements as well as the filtered services descriptions. (c) Finds the services according to the matched external descriptions.

6.2.2 Adding Semantics into ebRIM

Because the ebRIM information model enables catalogues to handle not only services but also other information resources such as data, it has been adopted by OGC. There have been efforts in the general information domain to add semantics to ebRIM (Dogac 2006; Dogac et al. 2005; Liu et al. 2005). The basic idea is to use those extensibility points such as new kinds of associations, classifications, and additional slots to record corresponding OWL classes, properties and related axioms such as subclassOf. However, few studies of registering OWL-S into ebRIM are available. Although OWL-S is mentioned by Dogac et al. (2004), only hierarchical OWL classes addressing service functionalities have been explored for registration in ebRIM. The semantics of service instances such as input and output cannot be used in the search. The system development here focuses on the geospatial domain and explores the registration of semantics for geospatial data, services, and service chains. An important characteristic of the geospatial domain is that an application often includes multiple modeling or processing steps involving large and heterogeneous data volumes. While the Dataset class is a core extension to the OGC-ebRIM profile, the ProcessModel class is a core extension here in that it addresses the analysis and knowledge sharing demands in the context of geospatial services.

6.3 Semantic Augmentation with Geospatial Catalogue Service

6.3.1 Semantics Registration in CSW

Extensions for registering semantics are created in the CSW-ebRIM profile. These extensions are designed to allow semantics-enhanced discovery and support on-demand delivery of geospatial data products. The following extensions shown as dark icons in Fig. 6.2 are made: (1) creating a new type of ExtrinsicObject, i.e. ProcessModel; (2) building new ClassificationScheme instances based on geospatial DataType and ServiceType ontologies; (3) adding slots to declare IOPE in the Service and ProcessModel classes.

(1) Creating ProcessModel

The semantics-enhanced catalogue service proposed in this book supports the discovery of process models. Both atomic services and service chains have process models that describe their behavior. A new association type DescribedBy, therefore, is defined with its sourceObject being a Service object and its targetObject being a ProcessModel object. The ebRIM model provides several standard classification schemes, such as ObjectType and AssociationType as a mechanism to provide extensible types. These classification schemes are called canonical classification schemes and can be extended by adding additional classification nodes. The ObjectType classification scheme defines the different types of RegistryObjects a registry may support, and therefore, the ProcessModel is defined as a classification node in this classification scheme, as shown in Table 6.1. The parent of the ProcessModel is a unique identifier referring to the parent classification node, namely ExtrinsicObject. The code of the ProcessModel contains a code that can be used in constructing the path. The path of the ProcessModel contains the canonical path from the root ClassificationScheme. The AssociationType classification scheme defines the types of associations between RegistryObjects. The

Table 6.1 The definition of ProcessModel in XML


```
<ClassificationScheme ...>
<ClassificationNode ...>
...
<ClassificationNode xmlns="urn:oasis:names:tc:ebxml-regrep:xsd:rim:3.0"
xmlns:dsig="http://www.w3.org/2000/09/xmldsig#" xmlns:xsi="http://www.w3.org/2001/
   XMLSchema-instance"
xsi:schemaLocation="urn:oasis:names:tc:ebxml-regrep:xsd:rim:3.0 http://laits.gmu.edu:8099/
   csw/schema/rim-3.0.xsd" id="urn:uuid:7755e34b-c794-4067-a409-7adb64bb6f7f"
   home="http://laits.gmu.edu:8099/csw/" objectType="urn:uuid:555c406c-2850-4b34-b75f-
   fe936f670960" status="Approved" parent="urn:uuid:6902675f-2f18-44b8-888b-
   c91db8b96b4d" code="ProcessModel" path="/ExtrinsicObject/ProcessModel">
<Name>
<LocalizedString xml:lang="en-US" charset="UTF-8"
value="ProcessModel"/>
</Name>
<Description>
<LocalizedString xml:lang="en-US" charset="UTF-8"
value="process model for the service "/>
</Description>
</ClassificationNode>
</ClassificationNode>
...
</ClassificationScheme>
```


association type DescribedBy is then defined as a classification node in the AssociationType classification scheme.

A Service instance can be either tightly coupled with a Dataset instance, or not associated with specific data instances, i.e. loosely coupled (ISO/TC211, 2005). In the tightly coupled case, the service metadata describes both the service and the geographic dataset, the latter being associated to the service using the association type OperatesOn. Figure 6.3 shows an example of this association. Loosely coupled services may have an association with DataTypes instead of specific data instances. This type of association is conveyed through the process model for the service. As shown in the Fig. 6.4, the input/output data slots in the process model can address loosely-coupled associations. If the registered service is actually a composite service (i.e. service chain), the composedOf slot in the ProcessModel can link to a detailed composite process model such as the OWL-S composite process.

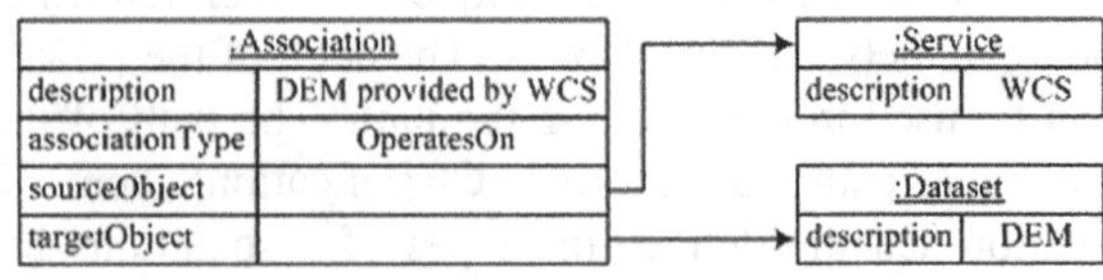

Fig. 6.3 Tightly coupled association between the service and data

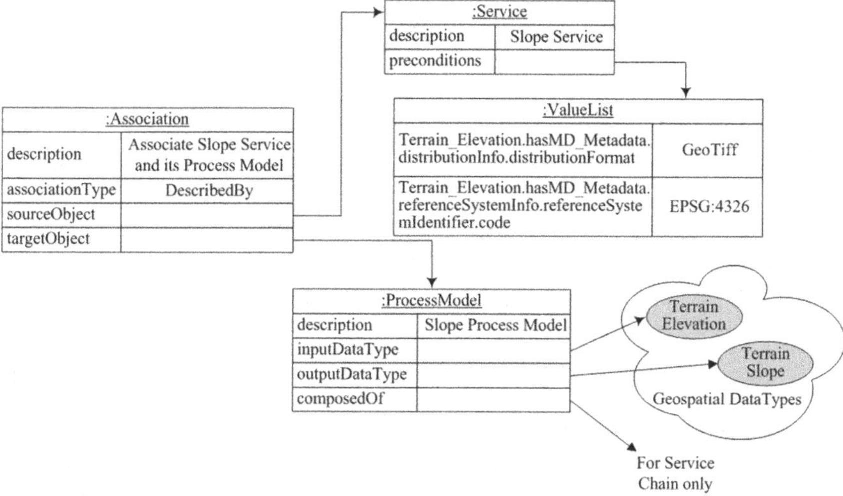

Fig. 6.4 An association between a service and its process model

(2) Building new ClassificationScheme instances

The ebXML Registry Profile for OWL proposed by the OASIS ebXML Registry Technical Committee (Dogac 2006) has provided a detailed guide on how to use ebRIM constructs to represent OWL constructs. An class in OWL must be mapped to a ClassificationNode in ebRIM. A classification scheme should be created for each ontology, and the classes belonging to this ontology should be represented as the classification nodes of this classification scheme. Therefore, two new ClassificationScheme instances as extensions are created, one for geospatial DataType ontology and the other one for geospatial ServiceType ontology.

Using these ClassificationSchemes, semantics can then be added by classifying geospatial data and services. Figure 6.5 shows that a dataset is classified according to the geospatial DataType classification scheme, using the associated classification node to specify its geospatial DataType. The lower part of Fig. 6.5 is an XML encoding example to illustrate this classification.

(3) Adding slots for IOPE

IOPE semantics for geospatial services were illustrated in Chap. 5. The input and output semantics for geospatial services address the loosely-coupled type association between services and data, and therefore are appropriate to be registered in the ProcessModel instances by adding inputDataType and outputDataType slots (Fig. 6.4). The values for these slots can be represented using ValueLists as shown in Table 6.2. Each value in the ValueList represents a unique identifier (e.g. URI) to the related geospatial DataType.

While input and output semantics address the loosely-coupled type association between services and data, the preconditions and effects for geospatial services are concerned more with instance association between services and data. For example, many individual services may be available under the slope ServiceType; however,

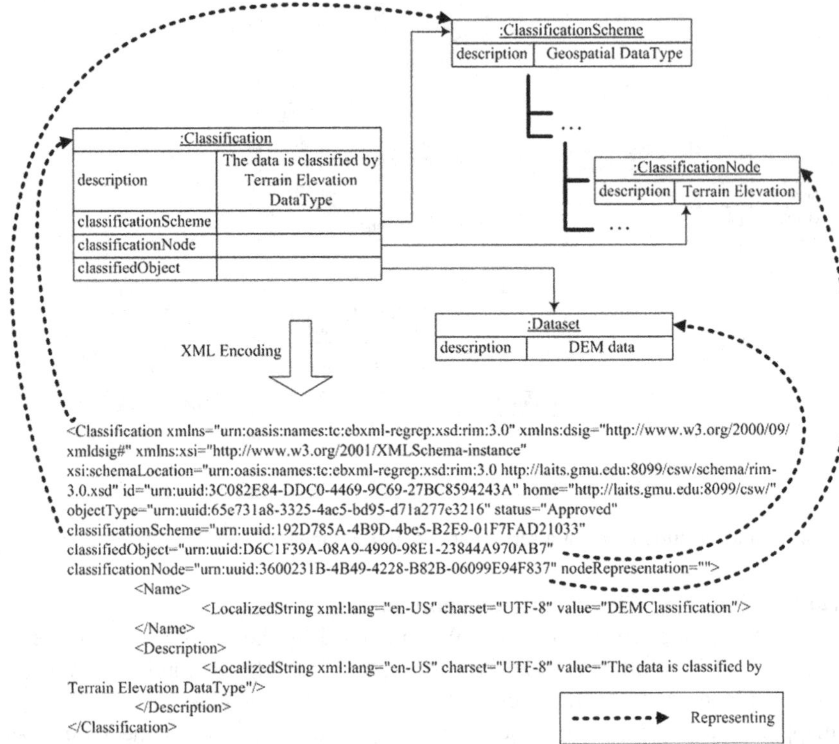

Fig. 6.5 Use of a geospatial DataType scheme for classification

Table 6.2 An example of inputDataType representation in XML

```
<Slot name="inputDataType" slotType="ProcessModel">
<ValueList>
<Value>http://www.laits.gmu.edu/geo/ontology/domain/GeoDataType.owl#
Terrain_Elevation</Value>
</ValueList>
</Slot>
```

each service may have its own metadata requirements for the input data, such as a particular file format or spatial projection. This often happens in the geospatial domain and is due to the complex nature of geospatial data, which are highly multidisciplinary and heterogeneous. Such an association differs from the tightly-coupled association addressed in the association type OperatesOn, because no specific dataset is associated. We may call it a mixed-coupling case. The metadata constraints specified in the preconditions can be valuable when searching a more specific service under a ServiceType. As shown in Fig. 6.4, a ValueList can specify preconditions, where each value in the ValueList represents a *contextual path*, a term proposed by Bowers and Ludäscher (2004), denoting a single concept, which may be in the context of one other concept through a series of properties. For example, "Terrain_Elevation.hasMD_Metadata.referenceSystemInfo.referenceSystemIdentifier.code" in Fig. 6.4 is such a path. Although its original purpose is to enable registration mappings and facilitate structural transformation of data, it does provide a way to identify a specific concept in a context, and thus can be used to identify a specific metadata element here. Figure 6.6 shows the mapping from a precondition represented using SPARQL to a contextual path.

6.3.2 Semantic Search Functionality

The extended catalogue contents and DL-based reasoning are used to formulate queries. Those extended catalogue contents are queried through the standard CSW interface. Table 6.3 shows a geospatial data query using the standard GetRecords operation. The classification nodes and scheme for geospatial DataTypes are used as a search condition in the query. TBOX reasoning is used to derive additional concepts as the search conditions in the query. For example, those classification nodes with subclass-superclass relations determined by hierarchical relationships in the ontology (i.e., subsumption reasoning in TBOX reasoning) are added to the query conditions to allow a more effective discovery.

Three types of match are defined including EXACT, SUBSUME, RELAXED. Let OntR denotes the requested concept and OntP denotes the provider concept, the three matching conditions can be expressed as the followings with the decreasing priority order:

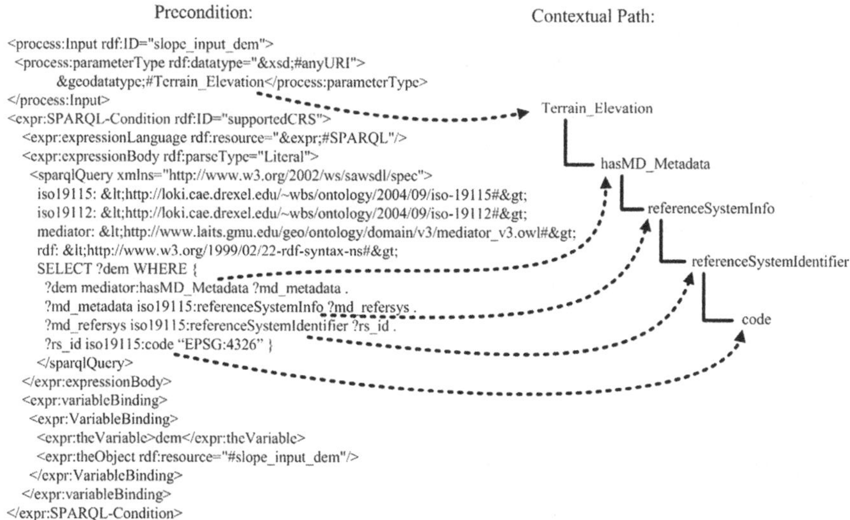

Fig. 6.6 Mapping from a SPARQL precondition to a contextual path

Table 6.3 A data query example, using geospatial DataType classification scheme

```
<?xml version="1.0" encoding="UTF-8"?>
<csw:GetRecords xmlns="http://www.opengis.net/cat/csw"
xmlns:csw="http://www.opengis.net/cat/csw" xmlns:ogc="http://www.opengis.net/ogc"
xmlns:gml="http://www.opengis.net/gml" version="2.0" outputFormat="text/xml"
   charset="UTF-8" outputSchema="http://www.opengis.net/cat/csw" startPosition="1"
   maxRecords="50">
<csw:Query typeNames="Dataset Classification ClassificationScheme ClassificationNode">
<csw:ElementSetName>full</csw:ElementSetName><csw:ElementName>/Dataset/
   </csw:ElementName>
<csw:Constraint version="1.0.0"><ogc:Filter><ogc:And>
<!–temporal condition–>
<ogc:PropertyIsGreaterThanOrEqualTo><ogc:PropertyName>/Dataset/beginDateTime
   </ogc:PropertyName>
<ogc:Literal>2005-01-10T00:00:00Z</ogc:Literal></ogc:PropertyIsGreaterThanOrEqualTo>
<ogc:PropertyIsLessThanOrEqualTo><ogc:PropertyName>/Dataset/endDateTime
   </ogc:PropertyName>
<ogc:Literal>2005-01-20T23:59:59Z</ogc:Literal></ogc:PropertyIsLessThanOrEqualTo>
<!–spatial condition–>
<ogc:BBOX><ogc:PropertyName>/Dataset/BBOX</ogc:PropertyName>
<gml:Box srsName="EPSG:4326">
<gml:coordinates>-122.2167,37.7994 -122.2167,37.7994</gml:coordinates></gml:Box>
   </ogc:BBOX>
<!–derived concept–>
<ogc:PropertyIsEqualTo><ogc:PropertyName>/Dataset/@id</ogc:PropertyName>
<ogc:PropertyName>/Classification/@classifiedObject</ogc:PropertyName>
   </ogc:PropertyIsEqualTo>
<ogc:PropertyIsEqualTo><ogc:PropertyName>/Classification/@classificationScheme
   </ogc:PropertyName>
<ogc:PropertyName>/ClassificationScheme/@id</ogc:PropertyName>
   </ogc:PropertyIsEqualTo>
<ogc:PropertyIsEqualTo>
<ogc:PropertyName>/ClassificationScheme/Description/LocalizedString/@value
   </ogc:PropertyName>
<ogc:Literal>geospatial data type</ogc:Literal></ogc:PropertyIsEqualTo>
<ogc:PropertyIsEqualTo><ogc:PropertyName>/Classification/@classificationNode
   </ogc:PropertyName>
<ogc:PropertyName>/ClassificationNode/@id</ogc:PropertyName></ogc:PropertyIsEqualTo>
<ogc:PropertyIsEqualTo><ogc:PropertyName>/ClassificationNode/@code
   </ogc:PropertyName>
<ogc:Literal>ETM_NDVI</ogc:Literal></ogc:PropertyIsEqualTo>
</ogc:And></ogc:Filter></csw:Constraint>
</csw:Query></csw:GetRecords>
```

EXACT: OntR equivalent to OntP
SUBSUME: OntP is a subclassOf OntR
RELAXED: OntR is a subclassOf OntP

If a user wants to find "Vegetation_Index" data, the query in Table 6.3 can be derived to search "ETM_NDVI" data that is semantically matched. As shown in Fig. 6.7, semantic middleware that can perform reasoning is created in front of the catalogue service, with no change to the legacy service interface. This semantic middleware is able to perform three types of discovery. The first is geospatial data discovery using a classification scheme for geospatial DataTypes. The query in the Table 6.3 is such an example. The Rodriguez and Egenhofer (2003) distance concept can be used as one option to control the enumeration of derived concepts. The distance is measured using the number of connected subclass-superclass arcs between two entity classes in the ontology, providing a reference value for assessing the similarity of entity classes.

The second type of discovery is service discovery. This kind of query includes discovering a service chain, since a service chain as a whole is a service. We have adopted the idea of a three-phase service discovery algorithm on UDDI by Sivashanmugam et al. (2003), except that we introduce the concept of the process

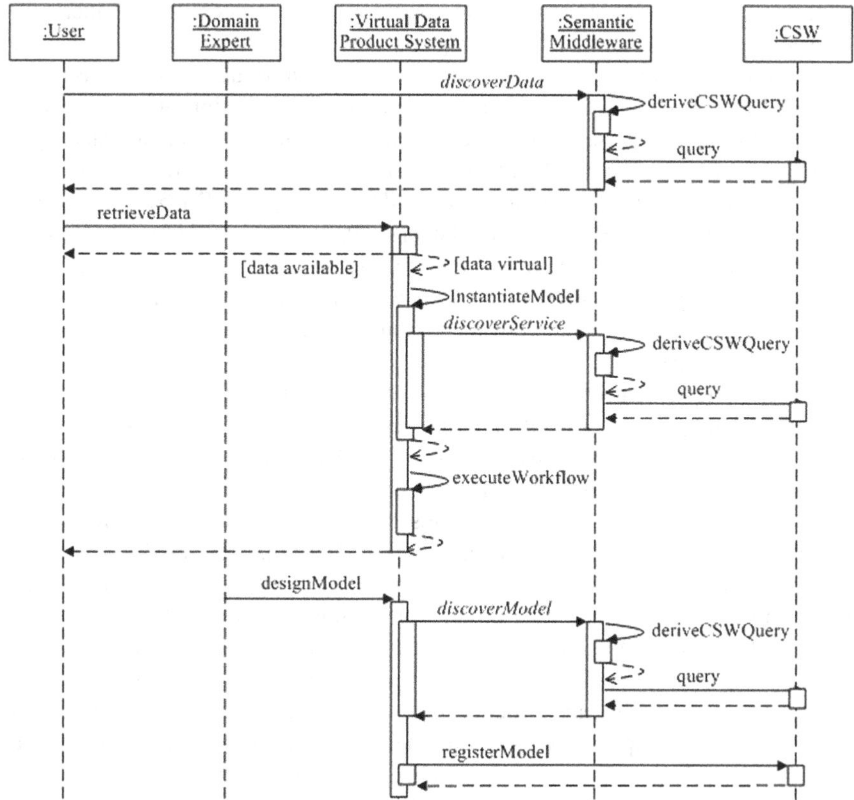

Fig. 6.7 UML sequence diagram illustrating the role of semantics-enhanced CSW in supporting virtual data production

model and use a different information model here. The users' requirements for the data and functional semantics of services are constructed as process templates using ontological concepts. A process template is defined as a tuple (F, I, O), where F is the semantic concept addressing the function of the process, I is a finite set of input semantic concepts and O is a finite set of output semantic concepts. In the first phase, process models are discovered using a geospatial ServiceType classification scheme. This is similar to geospatial data query. In the second phase, the process models resulting from the first phase are ranked in order of semantic similarity (Cardoso and Sheth 2003) between the input and output concepts of the selected models and the input and output concepts of the template, respectively. In the third optional phase, users' requirements on the execution semantics of services (here metadata requirements) are constructed using contextual paths. The third phase then involves service discovery using process models resulting from the second phase and optional context paths. The first and second phases can be combined to support the third type of discovery, the process model discovery. In addition, process model discovery can be flexible. For example, users might be interested in those models that can provide a certain output DataType. In this case, the output DataType serves as the only condition in the process model query.

It should be noted that the ebXML registry can use stored procedures to handle registered OWL semantics. For example, the path attribute in the ClassificationNode shown in Table 6.1 contains the canonical path leading from the parent nodes; therefore, it is feasible to derive semantically related geospatial DataTypes in the geospatial DataType classification scheme from this representation (e.g. path="/GeoDataType/.../Vegetation_Index/NDVI/ETM_NDVI") by using string comparison functions in the relational database. Predefined queries can be defined to invoke such stored procedures. However, use of stored procedures can only achieve limited reasoning functionality since the semantics in OWL can be fully explored only in its own syntax-aware reasoners due to its intrinsic logic nature (Dogac et al. 2005).

The semantics-enhanced catalogue service proposed in this book supports the discovery of not only geospatial data and services, but also the process models. When archived data or services are not available, it can find existing process models or automatically generate new process models, link process models to geoprocessing workflows or service chains through data and services discovery, and automatically execute service chains to provide virtual data products that can meet the original demands. The role of a semantics-enhanced geospatial catalogue service in supporting virtual data production is illustrated in the UML sequence diagram of Fig. 6.7. A virtual data product system that is capable of deliver geospatial information and knowledge on-demand can combine the discovery of geospatial data, services, and process models into three consecutive phases for automatic service composition: process modeling, process model instantiation, and workflow execution. In the process-modeling phase, the knowledge of a domain expert is captured through process models. The model design process can be manual or automatic. In manual model design, users can find existing process models, link different geospatial DataTypes and ServiceTypes together to create new process models, compose new process models from existing process models, or check whether existing process models are decomposable or not. To automate this design process, the ontology reasoning introduced in Sect. 6.3.2 and AI planning methods can be used to automatically generate new process models. In both manual and automatic model design, discovery of existing process models must be involved. When the process model design has been completed and evaluated through process model instantiation and workflow execution, the model can be registered in the catalogue for future use. In the process model instantiation phase, the process model can be bound to a concrete geoprocessing workflow or executable service chain through data and services discovery. A workflow execution engine, then, can use a service chain to generate on-demand data products. Through these three phases, a virtual data product is then materialized to a data instance. A detailed introduction is located in Chap. 7

Chapter 7
Automatic Composition of Geospatial Web Service

7.1 Geoprocessing Workflows, Geospatial Process Models, and Virtual Data Products

The geoprocessing algorithm provided by geospatial services may handle only a tiny part of the overall geoprocessing or may be a large aggregated processing. In both situations, the service should be well defined, have clear input and output requirements, and be independently executable. Such services can be chained to construct different *geoprocessing workflows* (or service chains)[1] for geospatial knowledge discovery. In a distributed data and information environment such as the World Wide Web, there are many independent data and service providers. A complex geoprocessing workflow may be scattered among multiple service providers. Therefore, standards for publishing, finding, binding, and execution of services are needed. By following the standards for interfaces, interoperability of different software systems is achieved. Web services developed by different organizations can then be combined to fulfill users' requests. Through the OWS testbeds, OGC has been developing a series of interface specifications under the OGC Abstract Service Architecture.

Figure 7.1 illustrates the relation among geoprocessing workflows, geospatial process models, and virtual data products. From the *knowledge discovery* perspective, the geoprocessing workflow transforms raw data into knowledge-added data products. For example, a landslide susceptibility data product, generated from the workflow processing the DEM data and Landsat ETM imagery, is a product of knowledge discovery. It has a process model that contains the landslide susceptibility, slope, aspect, land cover and NDVI computation subprocesses. In each of the subprocesses, it has its own model, i.e., calculating the landslide susceptibility index from the terrain slope and aspect, land cover type, and vegetation growing condition (i.e. NDVI) data, deriving the terrain slope and aspect from the DEM data, generating the land cover types using the image classification of the ETM

[1] Thereafter, in the context of this book we use the term "geoprocessing workflow" and "geoprocessing service chain" interchangeably.

P. Yue, *Semantic Web-based Intelligent Geospatial Web Services*, SpringerBriefs in Computer Science, DOI: 10.1007/978-1-4614-6809-7_7,

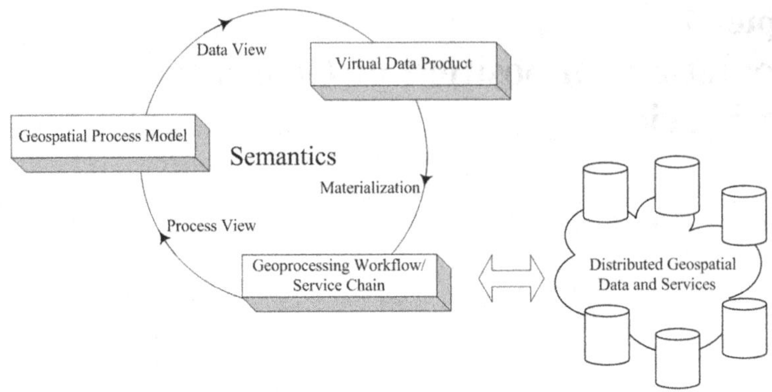

Fig. 7.1 Relation among geoprocessing workflows, geospatial process models, and virtual data products

imageries, and calculating ETM NDVI based on the NIR image (i.e. ETM Band 4) and red image (i.e. ETM Band 3). The *process model* of a geoprocessing workflow contains knowledge from a specific application domain. In a service-oriented environment, the generation of geospatial process models means generating an abstract composite process model consisting of the control flow and data flow among process nodes. The data flow focuses on the data exchange among process nodes, while the control flow concerns the order in which process nodes are executed. A process node represents one type of many individual services that share the same functional behaviors such as functionality, input, and output. Using a process model, users can produce a required data product even though the product does not really exist in any archive; therefore, a process model produces a *virtual data product*, comparable to the physically archived data products. The virtual data product represents a geospatial data type that the process model can produce, not an instance (an individual dataset). It can be materialized on-demand as an executable geoprocessing workflow or a service chain when all required geoprocessing methods and inputs, often discovered through a geospatial catalogue service, are available. By defining domain concepts to represent the semantics of geospatial Web resources (whether data, Web services, or service chains), the linkage among geospatial data, services, and geoprocessing service chains can be used for more effective discovery, automation, integration, and reuse in various applications.

7.2 "DataType"-Driven Automatic Service Composition

A geospatial question is concerning some kind of data, or more precisely, high level information or knowledge, for example, the landslide susceptibility data. Such high level information or knowledge is usually not directly available, especially for a specific location and time and thus some "service" is needed to derive them. Two simple computation models for landslide susceptibility index are assumed available as landslideSusceptibility services: one takes consideration of terrain slope and aspect, land cover type, and vegetation growing condition, and the other is based only on terrain slope and aspect. In each of the computation model, it also involves other models, such as deriving terrain slope from DEM and calculating NDVI as an indicator of vegetation growing condition. These other models may also involve more models. When both services and data/information/knowledge can be correctly described based on their thematic meanings and such descriptions are advertised in widely accessible catalogues, the answer to a particular geospatial question is potentially always available through reasoning on the thematic descriptions of data/information and recursively call related services for those data/information. This is a data-driven backward chaining process that creates an executable service chain starting with available input data and ending at the answer to a question.

In this approach for service chaining, the reasoning rules are primarily based on class hierarchical relationships defined in the service and data ontologies. According to semantic match priority discussed in Sect. 6.3.2, the preference order in a matching search for data and services is, in decreasing order, EXACT to SUBSUME to RELAXED.

The chaining process is based on service input–output concept matching. The request is a user-specified data product with metadata descriptions (e.g., spatial and temporal constraints), like the XML query described in Sect. 5.2.4. The system continually search regressively based on the match.

(1) If the match option is "EXACT", the data with exact-matched geospatial DataTypes are searched in CSW. If the match option is "SUBSUME", the data with exact-matched geospatial DataTypes are searched in CSW first. If such data is not available, then the data with subsume-matched geospatial DataTypes are searched in CSW. If the match option is "RELAXED", the data with exact-matched geospatial DataTypes are searched in CSW first. If such data is not available, then the data with subsume-matched geospatial DataTypes are searched. If the data is still not available, the data with relaxed-matched geospatial DataTypes are searched in CSW finally. This match strategy has two advantages: (a) high precision—those data with higher match degree are always got firstly; and (b) efficiency—the matched geospatial DataType collection are obtained through the one-time reasoning and then perform the keyword match successively. This is more efficient than the orderly match between the requested geospatial DataType and each geospatial DataType of available data, because in the later situation the reasoning process will repeat many times which will take lots of time in a large knowledge base.
(2) If matched data are not found in CSW, the system searches for associated geospatial ServiceTypes through the predefined "GeoDTSTAssociation" instances in the association ontology. Then, it constructs the CSW service query based on the geospatial ServiceType match option and the geospatial

ServiceType collection resulting from the Step 3. The same match strategy in the Step 1 is adopted here also.

(3) The system checks the available services according to the matching between the service output DataTypes and the requested geospatial DataType. If a matched service is found, the system builds the CSW data query according to the input geospatial DataType of the selected service, with those spatial and temporal constraints. This process continues recursively until all input data are available for the service chain. If finally some input data is not available, neither in the archives nor provided by the services, the chaining process will go back to an upper level, find another matched service, and repeat the above process again.

(4) When all binding data and services are available finally, the system converts them into an OWL-S Composite Process, and executes them to delivery the product to the requester.

7.3 Path Planning for Chaining Geospatial Web Services

A real world geospatial model is presented as the graph formulated using information from multiple geospatial semantic Web services. Nodes in the graph represent services and connectivity or edge weight is determined by the semantic matching of input and output of the services. The final optimum path is determined through path planning which consists of three interactive phases: path modeling, plan instantiation and service chain execution. The method presented here can be used to answer specific geospatial-related "what if" questions in a Web service environment.

A "path" is an ordered sequence of services that, when composed, can generate an executable service chain for problem solving. Thus the process of chaining geospatial Web services is a path planning process. A three-phase approach for the path planning is used. The first phase is to construct a logical model in which the most suitable service types are identified and logically connected. This is modeling phase. The second phase is to generate an executable service chain, a physical model, from the logical model through finding service instances of the chained service types. This is instantiation phase. The third phase is to actually execute the service chain.

7.3.1 Service Graph

Figure 7.2 shows a directed graph describing a partial landslide model. The nodes in this graph are services that will be needed to derive susceptibility index. The services are connected based on the semantic matches of their inputs and outputs, which are described by a geospatial DataType ontology defining the semantics of

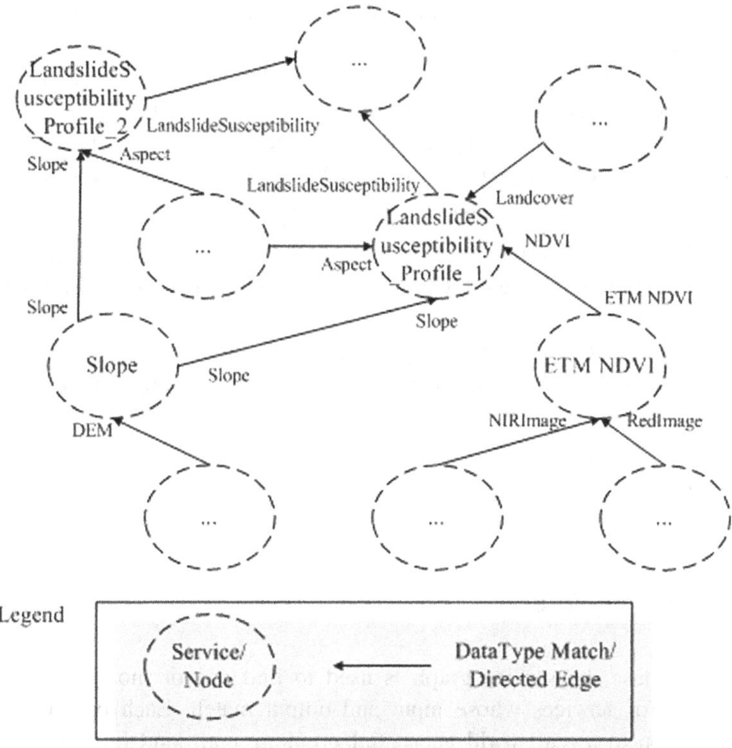

Fig. 7.2 A section of service graph

data and their hierarchical relationship. There are often multiple possible paths to research a specific node. For example, there are two landslideSusceptbility service nodes both generating landslide susceptibility index but taking different inputs. The connections between the services are assigned with positive weight values reflecting the different levels of the semantic matches. The weights are determined based on the hierarchical ontology relationships. Four levels of service matches are adopted: EXACT, SUBSUME, RELAXED and FAILED. To determine the connectivity from service *Node1* to service *Node2*, let *OntR* denote the input "DataType" of *Node2* and *OntP* denote the output "DataType" of *Node1*. The four levels of matching can be expressed as following with the increasing weight values:

EXACT: *OntR* equivalent to *OntP (Edge Weight Value=1)*
SUBSUME: *OntP* is a subclassOf *OntR (Edge Weight Value=2)*
RELAXED: *OntR* is a subclassOf *OntP (Edge Weight Value=3)*
FAILED: None of above matches. (*no connection or Edge Weight Value* $=+\infty$)

Information needed to generate the service graph is provided by the service profile of each service's OWL-S. The graph generated from service profiles is an abstract model which does not include information about physical availability of the involved services.

A section of graph is shown in Fig. 7.2. The graph is a directed graph. Let $TD(v)$ denote the node degree of node v which has m inputs, $IP = \{ip_1, ip_2, \ldots, ip_m\}$, and n outputs, $OP = \{op_1, op_2, \ldots, op_n\}$. For each input ip_i, the collection of services that can be potentially chained to v is given by $\sum_{j=1}^{i_l} S_j(ip_i)$, where i_l is the number of services the can provide semantically matched output for the ith input of node v. For each output op_i, the collection of services to which v can be chained is $\sum_{j=1}^{i_k} S_j(op_i)$, where i_k is the number of services whose inputs semantically match the ith output of node v.

Let $OD(v)$ and $ID(v)$ denotes, respectively, the outdegree and the indegree of node v. The node degree in the graph can be represented using Eq. (7.1).

$$TD(v) = OD(v) + ID(v) \tag{7.1}$$

where: $OD(v) = num\left(\sum_{i=1}^{n} \sum_{j=1}^{i_k} S_j(op_i)\right)$, $ID(v) = num\left(\sum_{i=1}^{m} \sum_{j=1}^{i_l} S_j(ip_i)\right)$.

7.3.2 Path Modeling

During this phase the service graph is used to find one or more sequences, or logical paths, of services whose input and output match. Each path provides a logical solution to a real world geospatial problem, e.g., landslide susceptibility data product in this case. The choice among various paths is subject to semantic control and various performance criteria. Semantic control includes both the correctness of a path and the degree of matching between connected services. Performance criteria are usually more important in the next phase—the plan instantiation, yet, in the current phase, it can still be used to help select a plan based on the length of the logical path. Multiple paths are usually constructed to provide alternative plans to deal with different instantiation and runtime possibilities. For example, if a required data or service is found to be not available or a service returns an error when executing a plan, the next suitable plan can be used. Considering a "what if" question, for example, "what is the landslide risk for location L at time T *if vegetation were changed*?", a logical path in Fig. 7.3 without the dotted rectangle is found.

7.3.3 Plan Instantiation

The instantiation process creates an executable service chain (physical path) by binding the service instances and available data to the logical path (i.e. plan). It consists of two steps: leaf node instantiation and service instance selection.

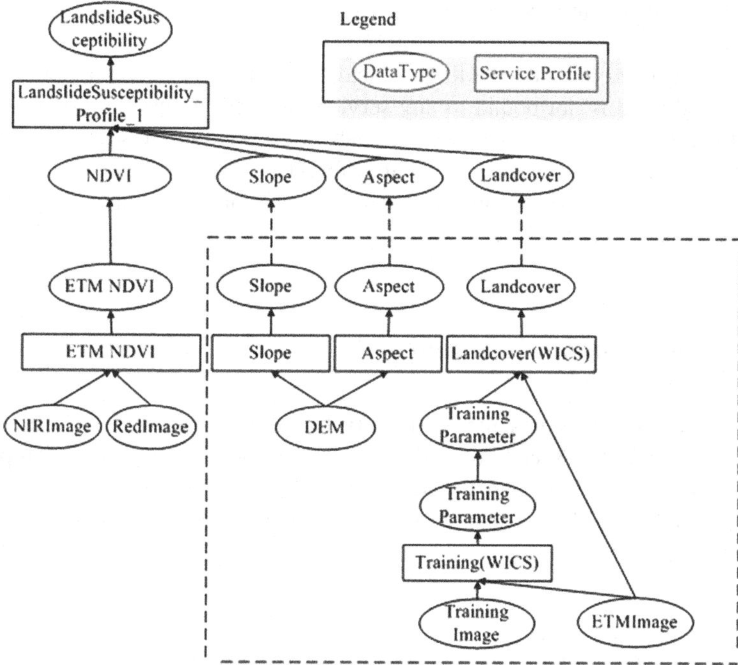

Fig. 7.3 An example of physical model

7.3.3.1 Leaf Node Instantiation

The concept of the "leaf node" is introduced to describe a service node for which at least one of its inputs, each described a specific "DataType" in the "DataType" ontology, is not connected to a service node in the logical path and thus data with correct DataType is needed for such an input. The process of binding a "Data-Type" to the available data is called leaf node instantiation. The available data may either be readily obtainable from some data provider or needs to be generated at run-time through a service chain. A geospatial catalogue is involved in this process to provide the information of data availability. In addition to the "Data-Type" constraint, more filtering requirements, such as spatial and temporal extents as instantiation parameters, are added to the query on the catalogue. If the requested data cannot be found, a matched service node in the graph can be selected to produce the requested data. Then the data query is moved on to the input "DataType" of the selected service node. The process continued until all input data are found available for the service chain. The resultant chain is called the "Physical Model". The process is exactly the "DataType"-driven service composition in Sect. 7.2. Figure 7.3 illustrates an example of the physical model resulting from the leaf node instantiation of an abstract model (i.e. logical path). The sub-chains inside the dotted rectangle represent the extension of the model after the instantiation process for a leaf node.

7.3.3.2 Service Instance Selection

Until now, each service node in the physical model is represented by an OWL-S service profile and is not bound to any service instance. A service profile can be bound to different service instances through corresponding service groundings. Different service instances are located at different physical addresses with related Quality of Service (QoS) information, such as network traffic and service performance. The selection of the service instance can be based on QoS information.

7.3.4 Service Chain Execution

The chaining result is represented as the OWL-S "Composite Process". It can be executed in an OWL-S engine. As mentioned in Sect. 5.2.5, the "Composite Process" of OWL-S can also be converted to any of the service composition languages such as BPEL to enable execution in the existing workflow engine for these languages.

7.4 Process Planning for Chaining Geospatial Web Services

Until now, the book has introduced two approaches for the automatic service chaining. However, in more sophisticated applications such as the wildfire prediction case, the human control as the decision support in the planning is more practical. This chapter addresses the semi-automatic geospatial service chaining through Semantic Web Service based process planning. The process planning includes three phases: process modeling, process model instantiation and workflow execution. The workflow execution phase is same as the service chain execution phase in Sect. 7.3.4. The details of process modeling and process model instantiation are addressed in the following using the wildfire prediction case.

There are numerous approaches for AI planning. Two main planning methods contribute to this work: regressive planning and hierarchical task network (HTN) planning. Regressive planning consists of backward state-space search (i.e. searching from the effects to the preconditions when considering an action), repeatedly simplifying the goal until the goal is achieved in the initial state (Brachman and Levesque 2003). The first action considered in the planner is the last one in the plan. Progressive planning, instead, is searching forward, i.e. from the preconditions to the effects. The goal in geospatial applications can be specified as a user-specified data product with metadata descriptions. Thus, the system can continually search regressively, using service input–output concept matching, until all input data are available for the service chain. HTN planning is more focused on task decomposition. Plans are refined by applying task decompositions. Each task decomposition reduces a high-level action to a partially ordered set of lower-level actions. Task decompositions, therefore, embody knowledge about how to implement actions. The key advantage of HTN is that, at each level of the hierarchy, a task is reduced to a small number of actions at the next lower level, so that the computational cost of finding the correct way to arrange those actions for the current task is small (Russel and Norvig 2003). Thus it reduces the complexity of reasoning by removing a great deal of uncertainty about the world.

7.4.1 Composite Process

In the DistanceBuffer process model, the Geocoder process, CTS process and Buffer process are chained together to provide the buffer based on the address. Figure 7.4 shows the semantic markup for these processes. The CTS can transform the projection of both geometry and coverage data, so its input and output are all represented by the "GeoDataType" concept. The output of the Geocoder process is annotated with the concept "GeocodedAddress", which is the conceptualization of "GeocodedAddressType" in the OGC OpenLS schema. The input and output of the Buffer process are represented as the "_GML" concept, which can represent any geometry type. If service chaining is based only on the match of the input–output concepts, then a SUBSUME match ("GeocodedAddress" is subClassOf "GeoDataType") is required for chaining the Geocoder and CTS, while a RELAXED match ("GeoDataType" is superClassOf "_GML") is required for CTS and Buffer. Much uncertainty remains when introducing the relaxed match because the parameter type in some services is too generalized. If the Distance-Buffer process model is defined as a composite process, the domain process knowledge contained in this process model can be captured, thus reducing the uncertainty caused by the reasoning to construct the process model among a large

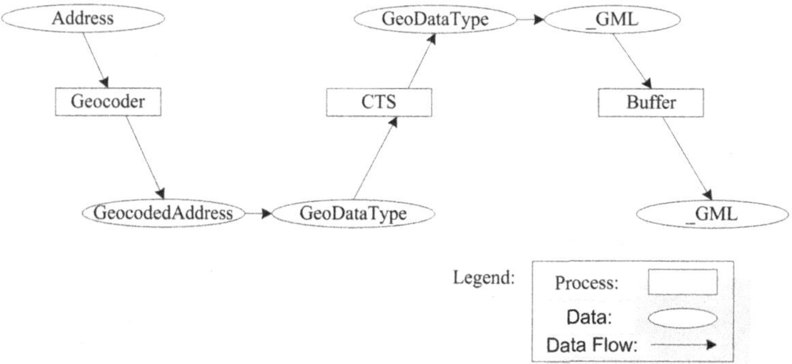

Fig. 7.4 DistanceBuffer process model

number of individual processes, just as the HTN planning does. In addition, a process library can be built by domain experts, and it can be reused and evolved.

The "Composite Process" ontology in the OWL-S is used to represent this kind of process model. A composite process can be characterized as a collection of subprocesses with control and data flow relationships. In OWL-S, the control flow is represented by the control constructs such as Sequence and Split. The data flow is specified through input/output bindings using a class such as ValueOf to state that the input to one subprocess should be the output of the previous one within a sequence.

Using the available composite processes, an abstract process model can be reduced to a structured set of subprocesses (perhaps further decomposed). The goal is to find a collection of atomic processes for some top-level composite process.

7.4.2 Process Modeling

A top-level process model can be built by the user using the "DataType" and "ServiceType" ontologies. In Fig. 7.5, John specifies in the top-level process model where a process to create the buffer, with "Buffer" as the "ServiceType", needs to take the "Address" as the input "DataType". Task decomposition is used in this phase. The DistanceBuffer composite process as a whole has the matched "ServiceType" and input "DataType". The match is based on the subsumption reasoning. Process templates are defined based on the data and functional semantics of services. A process template is defined as a tuple (F, I, O), where F is the semantic concept addressing the function of the process, I is a finite set of input semantic concepts and O is a finite set of output semantic concepts. The match process can be divided into two phases. The first is based on the concept match of functionality. It can reduce a large number of processes to a small set containing matched processes. The second phase finds the match of the input/output based on the result set of the first phase. An example of a lower-level process model is generated in Fig. 7.5.

7.4.3 Process Instantiation

Instantiation creates an executable service chain by binding the service instances and available data to the result of the process-modeling phase. It consists of two steps: physical model generation and validation.

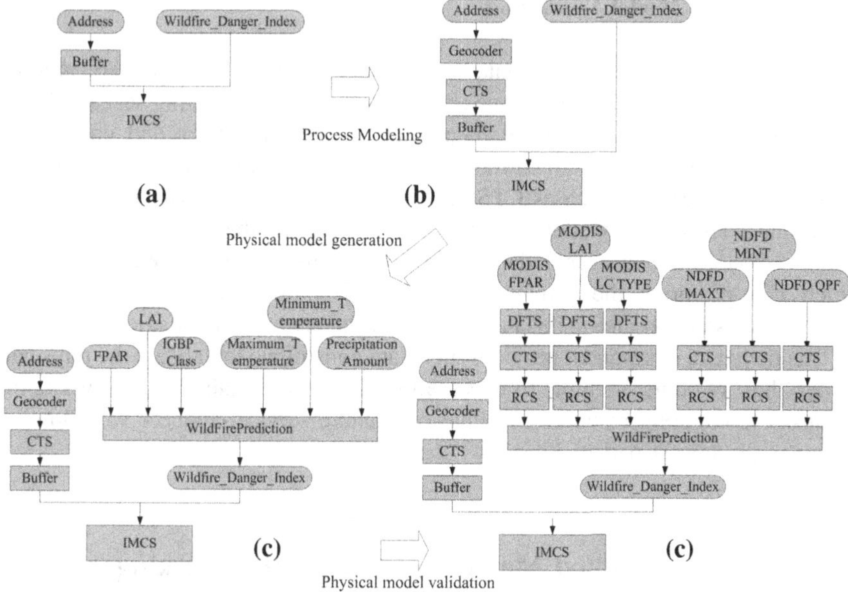

Fig. 7.5 Process planning for the wildfire prediction case

7.4.3.1 Physical Model Generation

The "WildFire_Danger_Index" data in the result from the process-modeling phase must be bound with the available data. The available data may either be readily obtainable from some data provider or needs to be generated at run-time through a service chain. A geospatial catalogue provides the information of data availability. In addition to the "DataType" constraint, more filtering requirements, such as spatial and temporal extents, are added to the query on the catalogue. If the requested data cannot be found, a process can be selected to produce the requested data. Then the data query is moved on to the input "DataType" of the selected process. If the selected process has the spatial constraints (e.g. wildfire prediction service), the spatial extent of the query for the input data of selected process should be adjusted correspondingly. The regressive planning process continued until all input data available for the service chain are found. Thus the "DataType"-driven service composition in Sect. 7.2 is used. The resultant chain is called the "Physical Model" (e.g. Fig. 7.5). Correspondingly, the model in the process-modeling phase is called the "logical model". In some simple cases such as creating a service chain to generate the slope data (firstly a WCS providing the DEM data, and then a slope calculation service generating the slope data), "DataType"-driven backward chaining is enough to derive a data product automatically without human intervention, and it can be characterized a method to enable opaque chaining, a type of service chaining defined in OGC Abstract Service architecture (Percivall 2002).

7.4.3.2 Physical Model Validation

In a physical model, the input data are available and subprocesses are atomic processes. However, in Earth science applications, many processing services have metadata constraints on the input data, such as the file format and coordinate reference system. As will be mentioned in the next section, the data reduction and transformation services such as CTS can modify the data to satisfy the metadata constraints. They are inserted automatically whenever the corresponding constraints are not satisfied. A metadata tracking component is introduced in Sect. 7.5, which can performs the semantic metadata generation, metadata validation, metadata satisfaction and metadata tracking in the service composition.

This metadata component provides metadata constraints to validate the physical model, thus avoiding attempts to execute invalid service chains and the waste of expensive computing resources. A final executable service chain or workflow is shown in the lowest part of Fig. 7.5.

7.5 Semantics-Enabled Metadata Generation, Tracking and Validation

When a composite process model is generated, the atomic process, which represents a set of services with same or similar functionalities, will now be grounded to an individual service instance based on the metadata constraints. In order to produce an executable service chain, the input of the process needs to be located using the detailed metadata, such as the file format, and coordinate reference system. There is already a significant literature identifying the usage of ontology-based metadata in the scientific workflow, most of them focusing on analyses of provenance data that are created from execution, rather than generation of input and output data description needed in the workflow before execution (Kim et al. 2006).

The generation of metadata before execution can bring several benefits to process model instantiation. First, it provides a context to the interpretation of data mining product in terms of data quality and reliability before the intensive execution of workflow, thus also contributes to the data provenance. Second, it helps to validate the concrete workflow with metadata constraints, thus avoids the execution of invalid service chain and the waste of expensive computing resources. And third, it helps to identify sub-chains whose execution results (the output) already exist, thus prevents the sub-chains from being executed.

This chapter explores the usage of semantic representation of the metadata for geospatial data that can be employed into the metadata constraint representation on the service chain and data product interpretation.

Based on the representation, a metadata tracking component has been implemented in the process-model instantiation phase. The component can perform semantic metadata generation, metadata validation and metadata tracking in the service chain. Based on this component, certain type of geospatial Web services, namely data reduction and transformation services such as data subsampling, reformatting, and reprojection, can be inserted automatically into the service chain to enable the metadata constraints satisfaction.

7.5.1 Metadata Constraints Specification

The metadata constraints are defined as metadata specifications that constrain the selection of instances of geospatial data and services during the materialization of process models into service chains. The scope of these constraints in the process model may vary. Some constraints apply to all the data and individual services involved in a process model (e.g. spatial area of interest). In this case, they are called *global metadata constraints*. Others may focus on a specific data set or service (e.g. file format supported); they are termed *local metadata constraints*. In a Semantic Web environment, all these constraints should refer to a semantics-enabled metadata representation. Therefore, an OWL description of geospatial metadata can be used to specify both global and local metadata constraints.

Both global and local metadata constraints can be represented using the ISO 19115 metadata ontology developed at Drexel University, USA (Drexel 2004). Global constraints are part of the users' goal (such as Table 5.6), the spatial and temporal constraints of the requested data product produced by the process model, e.g. a wildfire prediction for Bakersfield, CA on the next day.

Local constraints are the metadata constraints that the input data of an individual service must follow. Such constraints are represented as OWL-S preconditions. OWL-S preconditions can be presented using the SWRL, SPARQL, or other expression languages identified in the syntax of OWL-S. These preconditions are used for checking semantic consistency of services. The local constraints check is, therefore, equivalent to querying the knowledge base, which is a set of facts represented by the OWL, to check whether some condition for the input OWL individual is satisfied. ABOX reasoning is then used to infer implicit knowledge from the knowledge that is explicitly contained in the knowledge base, e.g. whether the input data file format satisfies the *unionOf* the OWL classes *GeoTIFF* and *NetCDF* (i.e., the format should be either *GeoTIFF* or *NetCDF*). The query of the OWL knowledge base for precondition checking makes SPARQL a more appropriate choice for precondition representation, since SPARQL is the W3C recommended standard query language for the RDF. Using the wildfire service as the example, the file format for some input data are specified using SPARQL, as shown in Table 5.4.

7.5.2 Semantic Metadata Generation and Propagation

Using a metadata catalogue service, the input data of the service chain, those that already physically exist in a data archive, can be queried to obtain detailed metadata information, using the global constraints as query filters. For example, the NOAA NDFD and NASA EOS MODIS data for input to the wildfire prediction service can be located using the temporal/spatial ranges identified in users' requests. If the metadata registered in the catalogue service does not have enough detail, a metadata generation component can be used to extract additional metadata from those data encoded in self-describing file formats such as HDF-EOS and GeoTIFF. The detailed metadata information for located data records in the catalogue service is transformed into OWL individuals and added into the OWL knowledge base to facilitate the precondition checking. Generation of metadata for intermediate and final data products depends on metadata propagation. Metadata propagation from the source input data to the final data products throughout the service chain depends on the metadata propagation on each atomic service.

In a service chain represented in Fig. 7.6, let *slope*(DEM) denote the output TerrainSlope, *ndvi*(ETMBand3, ETMBand4) denote the ETMNDVI, and *landslide*(TerrainSlope, ETMNDVI) denote the LandslideSusceptibility. The functional form representing this chain is

$$\text{LandslideSusceptibility} = \textit{landslide}(\textit{slope}(\text{DEM}),\ \textit{ndvi}(\text{ETMBand3},\ \text{ETMBand4}))$$

This functional representation is useful in analyzing metadata propagation. If we define a metadata propagation function for each atomic service, then the propagation of metadata through a service chain, called a metadata propagation model, can be represented as follows:

$$\begin{aligned}\text{MD}_{\text{LandslideSusceptibility}} = \textit{func\_landslide}(& \\ & \textit{func\_slope}(\text{MD}_{\text{DEM}}), \\ & \quad \textit{func\_ndvi}(\text{MD}_{\text{ETMBand3}},\ \text{MD}_{\text{ETMBand4}}))\end{aligned}$$

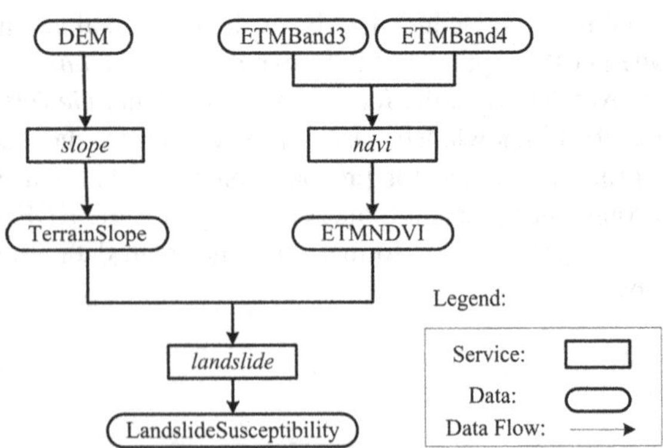

Fig. 7.6 Graphic representation of a service chain

where MD is the metadata description and the *func*'s are a set of metadata propagation functions that modify the MD appropriately for each atomic service in the service chain. A metadata propagation function for each service modifies the metadata for the input data. The modified metadata is then passed to the output data. This functional representation helps in understanding metadata propagation. The metadata of the final data product can be described solely in terms of the source input data and a set of metadata propagation functions. The functional representation of the metadata propagation model described above also implies that, when metadata for source input data is generated from the metadata catalogue service, the metadata propagation functions need only be defined appropriately in order to derive the metadata for a final data product.

The metadata propagation functions themselves must be tailored to specific geoprocessing services and particular metadata elements. For the purposes of metadata propagation, it is useful to identify two types of metadata propagation as shown in Fig. 7.7: unary and *n*-ary functions. A unary function has one input data set, and outputs the requested data product. An *n*-ary function takes *n* inputs to output the requested data product (where $n > 1$).

It is assumed that when a service processes input data, values for explicitly described metadata elements can be changed while values of other metadata elements are transferred unchanged to the data output by the service. Explicitly described metadata elements can be specified in the execution semantics of geospatial services, i.e. using OWL-S preconditions/effects. In OWL-S, the effects use the OWL Expression class for representing their values, which is the same as preconditions do. Thus, the processing of both preconditions and effects is similar, except that the metadata in the Expression of effects is processed as the update of the metadata for the output data product. This assumption for metadata propagation is reasonable in real geospatial Web service applications. For example, a slope computation service changes only the thematic meaning of data and a data format transformation service changes only the file format of the data. For the unary case, except for the updated metadata, the metadata of the output data is the same as the metadata of the input data set. In the *n*-ary case, except for the updated metadata,

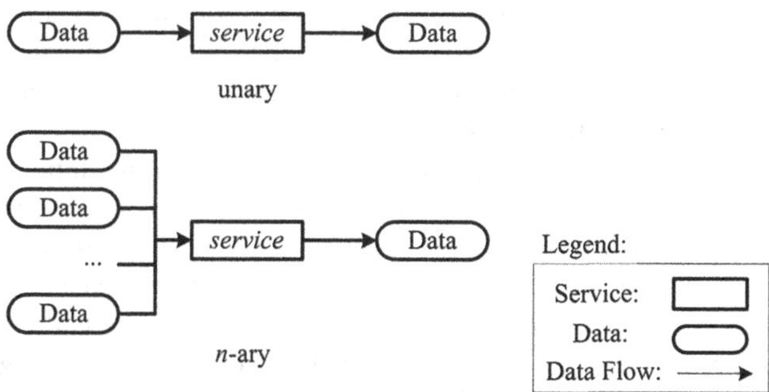

Fig. 7.7 Types of metadata propagation

the metadata of the output data is the same as the metadata of the principal input data set. The principal input is identified using the data model of the output and specified as the first input of the process in the OWL-S. For example, assume a geospatial expert wants to know the possibility of having wildfire(s) within a 300 km radius of interested area. An image cutting process, which uses an input polygon to cut the image, creating an image containing the values of the desired area only, is used. The input for image data serves as the principal input data since it uses the raster data model, the same as the output does. If there are multiple inputs using the raster data model, any input among them can serve as the principal input, because these inputs share some common characteristics after coregistration using the data reduction and transformation services.

7.5.3 Metadata Constraints Satisfaction

Compliance with the global constraints is validated by setting query filters when locating the input data of the service chain from the metadata catalogue service. The local constraints are validated through the OWL-S precondition checking mentioned in Sect. 7.5.1. When local constraints are not satisfied, an automatic constraint satisfaction strategy can be employed to modify the service chain. In the Earth science domain, data reduction and transformation services such as coordinate transformation service, data format transformation service are common to most geospatial analysis, data mining, and feature extraction applications. They can modify the data to satisfy the metadata constraints. The rule for using these services applies for all geospatial users, i.e., they can be used whenever the corresponding metadata constraints are not satisfied.

The use of only an individual data reduction and transformation service to satisfy a particular metadata constraint is simple and requires only the insertion of this service before the constrained service. However, when multiple data inputs and multiple metadata constraints are involved, this problem becomes complex due to the possible interactions among these services and the context sensitivity of applications.

The flow diagram of Fig. 7.8 shows the process for automatically validating that geospatial metadata constraints are satisfied. It is made up of two loops. The outer loop is controlled by data input to the current service. For a given data input, the inner loop performs the precondition check (i.e., validating that metadata constraints are satisfied) for each precondition constraining this input.

When the precondition check fails, an appropriate data reduction and transformation service is inserted into the service chain. The insertion of the data reduction and transformation service to satisfy certain metadata constraints is implemented as a domain control procedure (e.g., some source code in the computer software program). The domain knowledge on the common usage of these data reduction and transformation services, therefore, is embodied in these procedures. To make the proposed flow work, preconditions for each geoprocessing

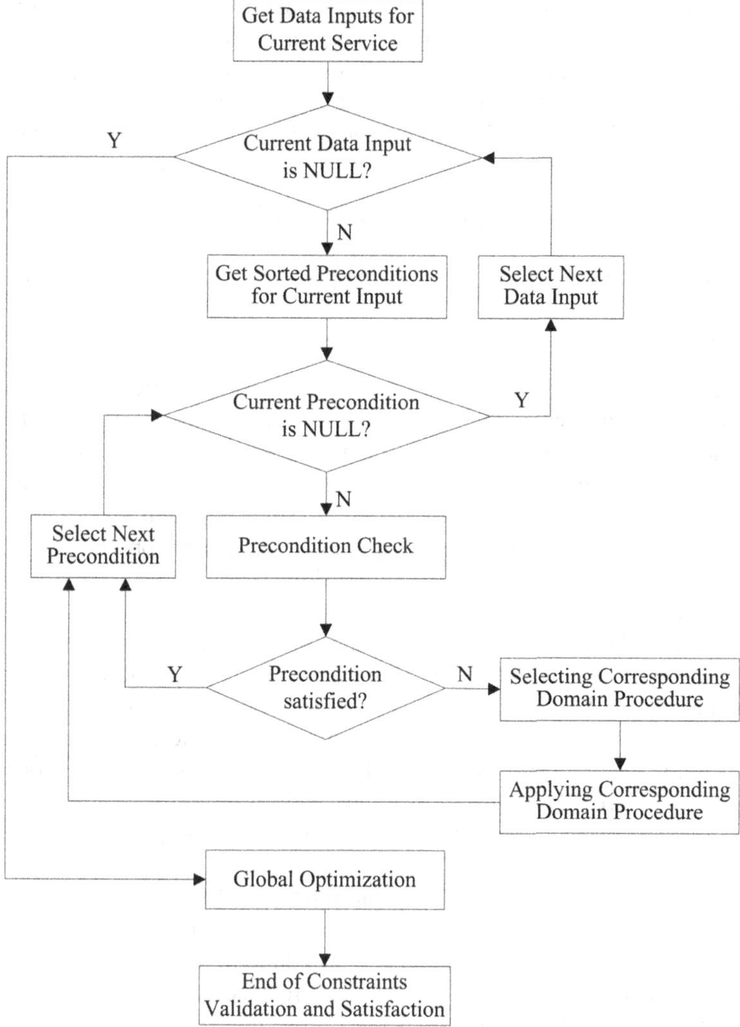

Fig. 7.8 Flow diagram illustrating automatic geospatial metadata constraints validation and satisfaction

service must be defined modularly. Each precondition should make only one type of constraint on one input parameter. Such a requirement facilitates the identification of a certain metadata constraint by examining each precondition, which can then help select the corresponding procedure if the specific constraint is violated. For example, two preconditions for the input data of the slope service are defined: one is for the coordinate reference system, and the other is for the file format.

Since the definition of preconditions is based on the ISO 19115 ontology, the common procedure is to identify the constraint and extract the information from the precondition (i.e., template queries can be defined to check preconditions). Figure 7.9a shows an example, where the constraint on projection is not satisfied in the top part of the figure. Figure 7.9b shows how to extract the projection code from the spatial projection precondition and then transfer it to the target projection parameter of the Coordinate Transformation Service. The procedure includes the following steps:

(1) Transforming a precondition into an RDF graph, which represents a knowledge base consisting of facts;
(2) Defining a template query to extract the parameter value from the knowledge base;
(3) Assigning the data flow in the updated service chain.

Figure 7.9b shows an RDF graph generated using a precondition in the SPARQL expression. Template queries such as SPARQL queries can be defined to extract the parameter values required in the modified service chain, as represented in the data flow of an OWL-S composite process shown in Fig. 7.9b. Thus each domain procedure modifies not only the control flow (through service insertion) but also the data flow (through input/output bindings). After applying the domain procedure, the metadata for input data of the slope service is updated as shown in the lower part of Fig. 7.9a.

The order of the preconditions, in the inner loop of the process shown in the Fig. 7.8, affects the order of services inserted. Figure 7.10 shows the services inserted between a Web coverage service and slope computation service when processing from the spatial projection precondition and file format precondition respectively. In this use case, the coordinate transformation service can process the data only in the HDF-EOS file format. The indexes in Fig. 7.10 show the processing flow step by step during the employment of domain procedures. The first chaining result of the two (Figs. 7.10a, b) is preferred since it has fewer computational steps. Therefore, the inner loop can start from different preconditions and adopt the shortest path method, favoring conclusions resulting from shorter paths (i.e. fewer services) of the service chain. In addition, when getting the sorted preconditions before the inner loop, some preferences can be imposed on the domain procedures. For example, the spatial projection precondition has a higher priority than the resolution precondition because the regridding operation in the resolution conversion service must be executed in the correct coordinate reference system to meet the requirements of a fire prediction service.

The service chain in Fig. 7.10b can be optimized to improve the execution efficiency. When two services transforming between data formats are joined sequentially, they can be replaced by a new service transforming between data formats, with its inputs those of the first service and output that of the second service. With metadata generated for intermediate data products, it is possible to use it as filters to generate queries to a metadata catalogue for data that will prevent those sub-chains from being unnecessarily executed. Therefore, as shown in Fig. 7.8, when precondition checking and action for all inputs are finished, global optimization can be employed. It consists of two steps: identifying

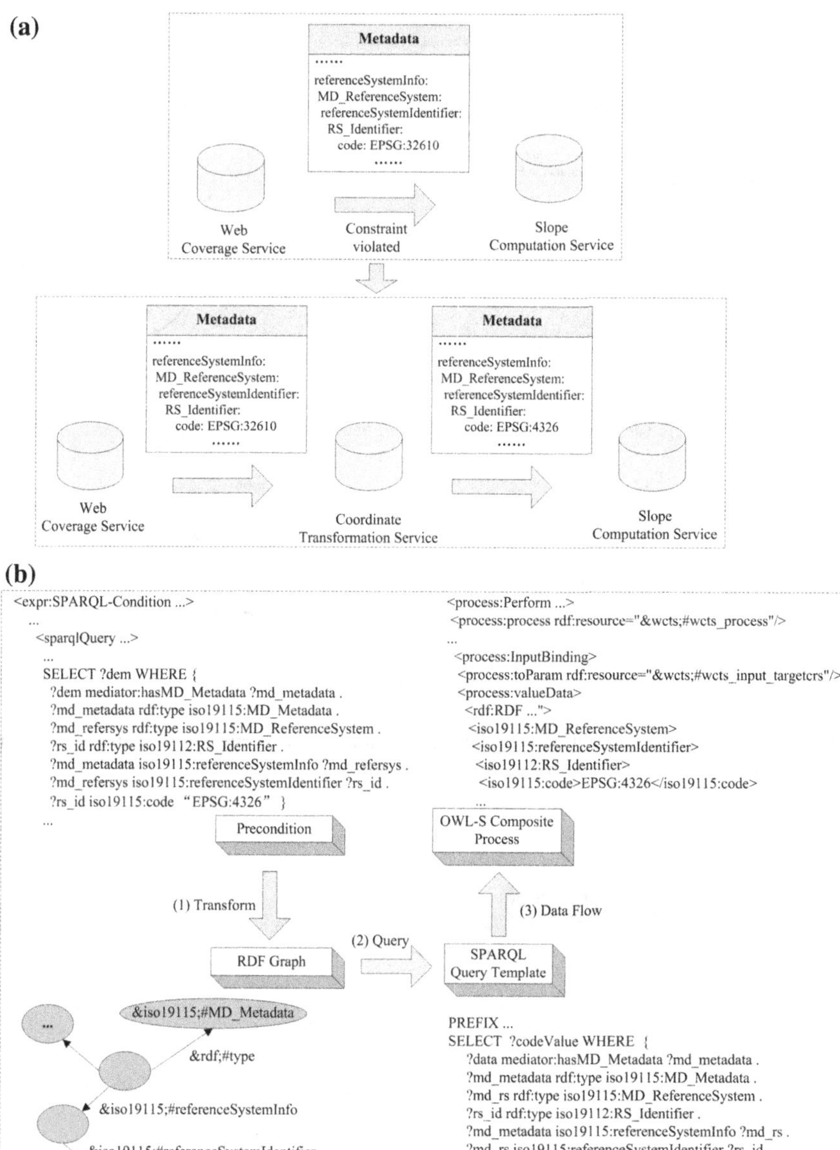

Fig. 7.9 Data flow in domain procedures: **a** metadata update after applying the domain procedure; and **b** steps included in the domain procedure

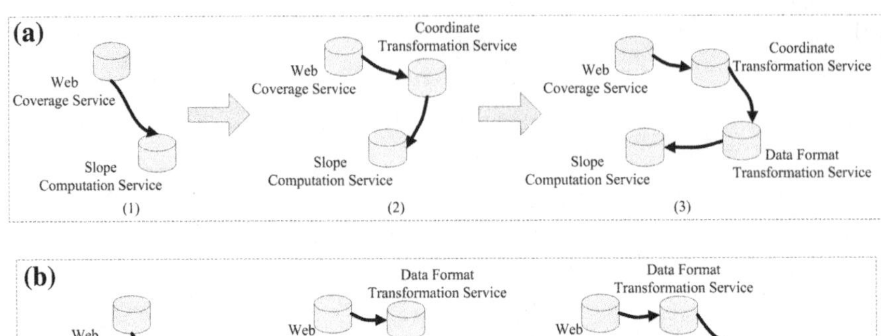

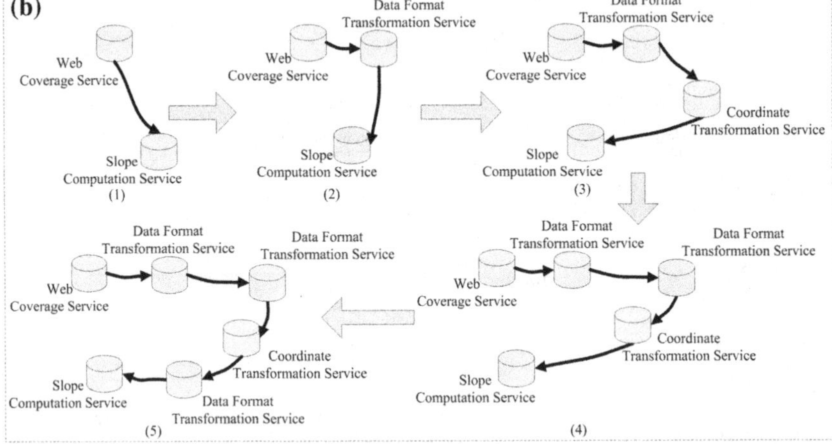

Fig. 7.10 Precondition satisfaction for multiple preconditions: **a** starting from spatial projection precondition; and **b** starting from file format precondition

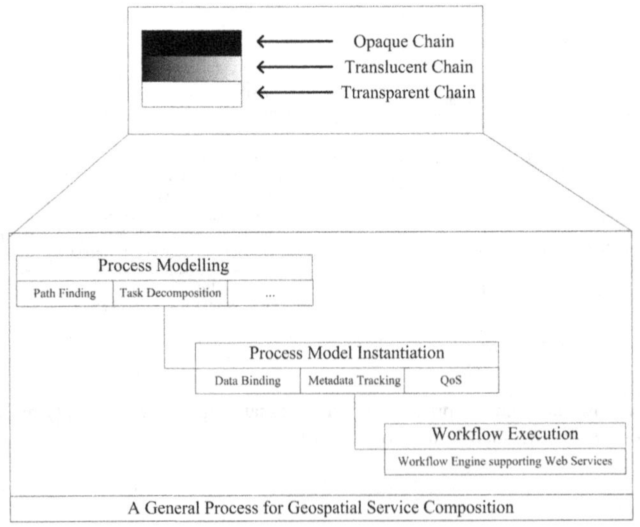

Fig. 7.11 General process for geospatial service composition

sub-chains whose outputs already exist, to prevent repeated execution through reorganizing the service chain, and decreasing service redundancy by not chaining two services of the same functional type successively.

7.6 Summary

OGC Abstract Service architecture (Percivall 2002) identifies three types of service chaining:

(1) User-defined (transparent)—the human user defines and manages the chain.
(2) Workflow-managed (translucent)—the human user invokes a service that manages and controls the chain. The user is aware of the individual services in the chain.
(3) Aggregate (opaque)—the human user invokes a service that carries out the chain. The user has no awareness of the individual services in the chain.

Through approaches in this book, it is possible to address all of them. Figure 7.11 shows a general process for geospatial service composition. "Data-Type"-driven service composition can be characterized as the "Data Binding" step in the generation of "physical model". The bound data can be either an archived data, or a "virtual data product" which is the output of some service chain. The "logical path" and "physical path" in the path planning are equivalent to the "logical model" and "physical model" in Fig. 7.11. The "DataType"-driven service chaining and path planning can be identified as the opaque chaining. However, in more sophisticated application such as the wildfire prediction case, human control as the decision support on the generation of logical models is more practical and can reduce the uncertainty in the automatic service composition. From this perspective, the process planning approach can be characterized as the translucent chaining. Users can also chain the multiple services manually by themselves and go the third phase (i.e. workflow execution) directly. This can be characterized as the transparent chaining. These different types of chaining methods can be combined with semantic catalogues in Chap. 6 to support on-demand delivery of geospatial information and knowledge and provide virtual data products.

Chapter 8
Prototype Implementation and Result Analysis

A prototype is implemented as part of the work in this book. The OWL-S Application Programming Interface (API) (MINDSWAP 2004) is used for OWL-S parsing and execution. The approach has been implemented in a common data and service environment enabled by the OGC and W3C standards. Section 8.1 introduces the extensions to OWL-S and OWL-S API. Section 8.2 presents the prototype architecture and its implementation. Finally, Section 8.3 provides analysis of results.

8.1 Software

Currently, OGC Web services are not equivalent to the W3C SOAP-based Web services. Most OGC Web service implementations provide access via HTTP GET and HTTP POST. They do not support SOAP. Since WSDL can describe the HTTP GET/POST bindings in addition to the SOAP binding, the HTTP GET and POST bindings can still be supported in the service grounding. Figure 8.1 and 8.2 show some part of WSDL description for WCS and WPS respectively. However, the WSDL grounding in OWL-S cannot handle the mapping of the multiple OWL-S inputs into a complex WSDL schema type in a message. Following the XML Message handling in BPEL, the WSDL grounding is extended by the additional property "wsdlMessagePartElement" which contains the XPATH (Clark and DeRose 1999) to locate the certain element in the complex type. Table 8.1 shows a snippet of WSDL and service grounding for a WPS buffer process.

OWL-S API provides a Java API for programmatic access to read, execute and write OWL-S service descriptions. The API provides an ExecutionEngine that can invoke AtomicProcesses that has WSDL groundings, and CompositeProcecesses that uses control constructs such as Sequence, and Split-Join. It has been extended in this work to support the HTTP GET and POST invocation in addition to the SOAP invocation it already has. The most advanced version of OWL-S that

P. Yue, *Semantic Web-based Intelligent Geospatial Web Services*,
SpringerBriefs in Computer Science, DOI: 10.1007/978-1-4614-6809-7_8,

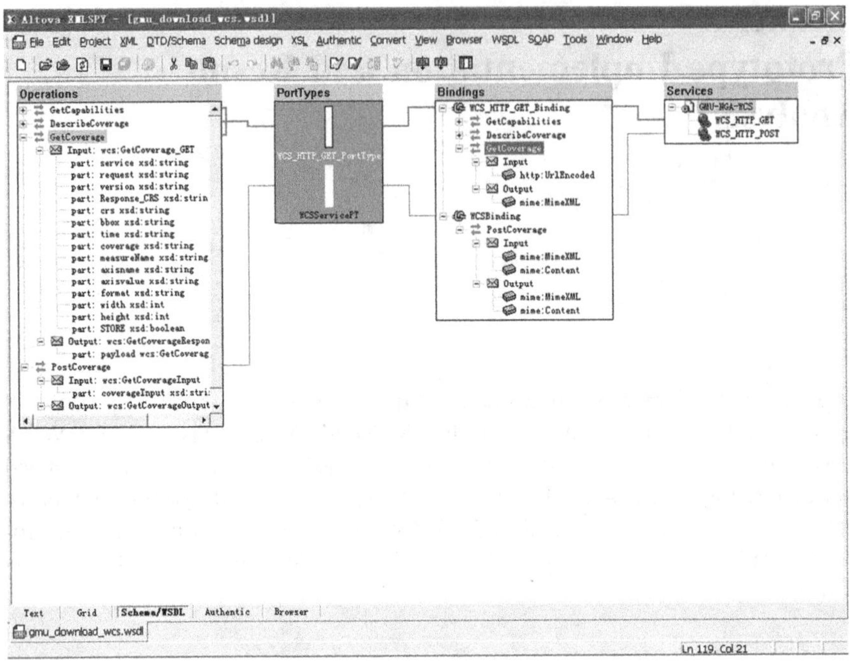

Fig. 8.1 WSDL for WCS

OWL-S API supports currently is version 1.1. It has been extended to also support some new features in the pre-release version of 1.2, including support of the SPARQL precondition. OWL-S API is implemented on top of the Jena (HP 2006), a Java framework for building Semantic Web applications. Jena has provided a programmatic environment for RDF, OWL, and SPARQL, and includes a rule-based inference engine. By using Jena, one can parse, create and search the concepts in semantic models based on RDF technique. Both TBOX and ABOX reasoning are supported by reasoners in Jena.

8.2 Prototype Implementation

OWLSManager, a system for the management of OWL-S files that can deploy and undeploy OWL-S files into the knowledge base, is developed It is composed of the following components (Fig. 8.3).

Client. Assists users in formulating goals on the basis of the ontology supported by the system. Users can also use it to contribute a composite process or atomic process of OWL-S into the knowledge base.

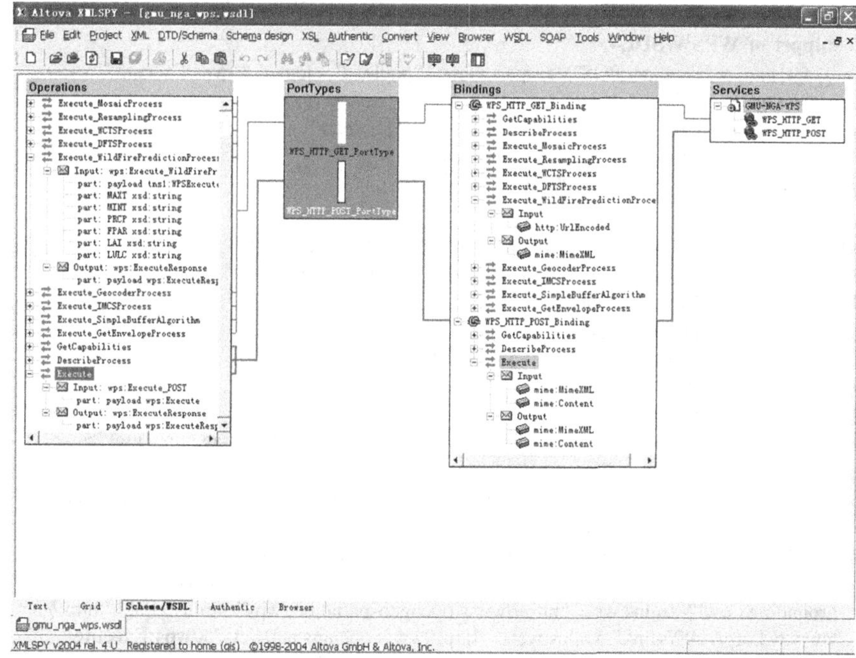

Fig. 8.2 WSDL for WPS

Knowledge Base. Includes the definitions of geospatial domain ontology and services ontologies. An inference engine is attached for reasoning.

Plan Generation. Uses AI planning approaches to generate the process model to achieve the users' goals, i.e. data products. The metadata tracking component is used to validate the process model and generate the executable service chain. The final result is produced as the OWL-S composite process and sent to the chain execution component for execution.

Chain Execution. Executes an OWL-S composite process by invoking each individual service and passing the data between the services according to the flow specified in the composite process. The individual services are invoked using the service groundings. The composite process can also be converted to a workflow language and executed in the corresponding workflow engine.

Catalogue. Provides a mapping description between ontologies and the CSW registration information model in order to register ontologies into the CSW to facilitate discovery of data and services.

OWL is used as the language for representing geospatial semantics. Jena Transitive and the OWL-Micro Reasoners (HP 2006) are selected for reasoning. The first one is preferred because of its efficient TBOX reasoning. However it can

Table 8.1 A snippet of WSDL and service grounding for the WPS buffer process

```
<!–snippet of WPS WSDL–>
<message name="Execute_POST"><part name="payload" element="wps:Execute"/>
  </message>
<message name="ExecuteResponse"><part name="payload"
  element="wps:ExecuteResponse"/></message>
<portType name="WPS_HTTP_POST_PortType">
...
<operation name="Execute"><input message="wps:Execute_POST"/>
<output message="wps:ExecuteResponse"/></operation></portType>
<!–snippet of service grounding–>
<grounding:wsdlInputMessage
  rdf:datatype="&xsd;#anyURI">&wps_buffer_wsdl;#Execute_POST
  </grounding:wsdlInputMessage>
<grounding:wsdlInput>
<grounding:WsdlInputMessageMap rdf:ID="wps_buffer_wsdlinputmessagemap_gml">
<grounding:owlsParameter rdf:resource="&buffer_profile;#buffer_input_gml"/>
<grounding:wsdlMessagePart rdf:datatype="&xsd;#anyURI">&wps_buffer_wsdl;#payload
  </grounding:wsdlMessagePart>
<groundingx:wsdlMessagePartElement rdf:datatype="&xsd;#string"><![CDATA[
<context type="xpath" xmlns="http://www.laits.gmu.edu/geo/ontology/domain/
  groundingx.owl" xmlns:wps="http://www.opengeospatial.net/wps" xmlns:xlink="http://
  www.w3.org/1999/xlink" xmlns:wcts="http://www.opengis.net/wcts">/wps:Execute/
  wps:DataInputs/wps:Input[position()=1]/wps:ComplexValue</context>]]>
  </groundingx:wsdlMessagePartElement>
<grounding:xsltTransformationString><![CDATA[
<xsl:stylesheet version="2.0" xmlns:xsl="http://www.w3.org/1999/XSL/Transform"
  xmlns:rdf="http://www.w3.org/1999/02/22-rdf-syntax-ns#" xmlns:mediator="http://
  www.laits.gmu.edu/geo/ontology/domain/v3/mediator_v3.owl#" xmlns:iso19115="http://
  loki.cae.drexel.edu/~wbs/ontology/2004/09/iso-19115#" xmlns:gml-ont="http://
  loki.cae.drexel.edu/~wbs/ontology/2004/09/ogc-gml#" xmlns:ows="http://
  www.opengeospatial.net/ows" xmlns:xlink="http://www.w3.org/1999/xlink"
  xmlns:geodatatype="http://www.laits.gmu.edu/geo/ontology/domain/GeoDataType.owl#"
  xmlns="http://www.opengis.net/gml">
<xsl:import href="http://www.laits.gmu.edu/geo/ontology/owls/xslt/owl2gmlpacket.xsl"/>
</xsl:stylesheet>
]]></grounding:xsltTransformationString>
</grounding:WsdlInputMessageMap></grounding:wsdlInput>
<grounding:wsdlInput>
```

not support the ABOX reasoning in the precondition checking. The second is used for the OWL-S precondition check.

Using the guidelines of the ebRIM profile for CSW, the CSW implementation, developed and maintained by Laboratory for Advanced Information Technology and Standards (LAITS) from George Mason University (Wei et al. 2005), has extended ebRIM using international geographic standards: ISO 19115 Geographic Information—Metadata (including part 2: Extensions for imagery and gridded

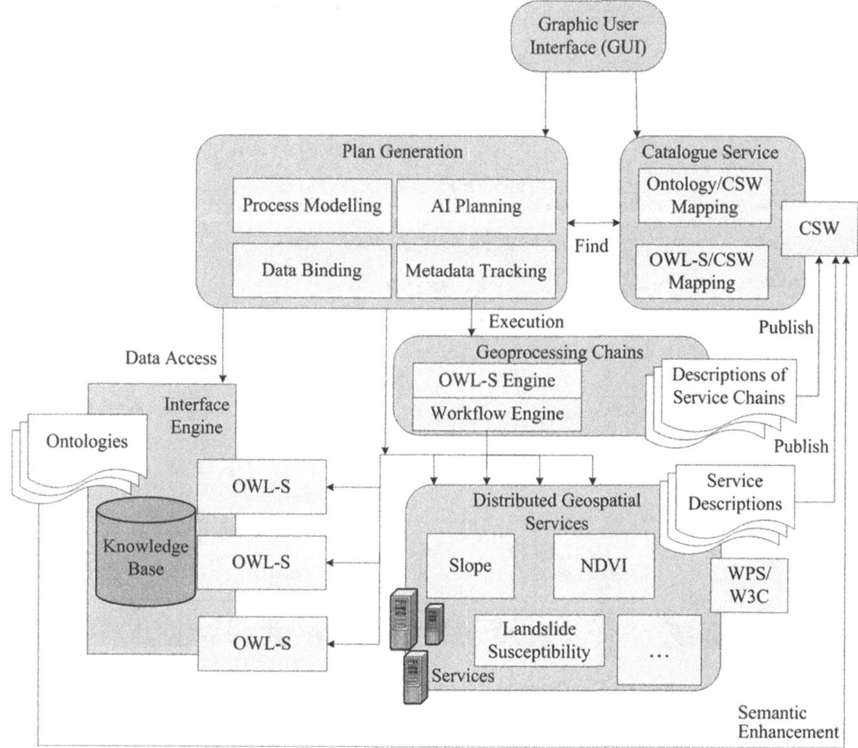

Fig. 8.3 System architecture

data) and ISO 19119 Geographic Information—Services. The ebRIM is extended with ISO 19115 and ISO 19119 in two ways. The first is by importing new classes into the ebRIM class tree, deriving new metadata classes from existing ebRIM classes. The new Dataset class is used to describe geographic datasets. Many new attributes are added to the Dataset class based on ISO 19115 and its part 2. The second way to extend ebRIM is to use Slots to extend an existing class. The Service class included in ebRIM can be used to describe geographic services, but the available attributes in the class Service are not sufficient to describe geospatial Web services. New attributes derived from ISO 19119 are added to the Service class through Slots. The geospatial semantics are registered in the CSW and can be queried through the CSW interface.

OWLSManager provides a client interface to the user to perform four types of primary functions:

(1) The OWL-S Files Management functions:

   a. Set Schema (geospatial semantics schema): Provides the knowledge base for the reasoner (Fig. 8.4).

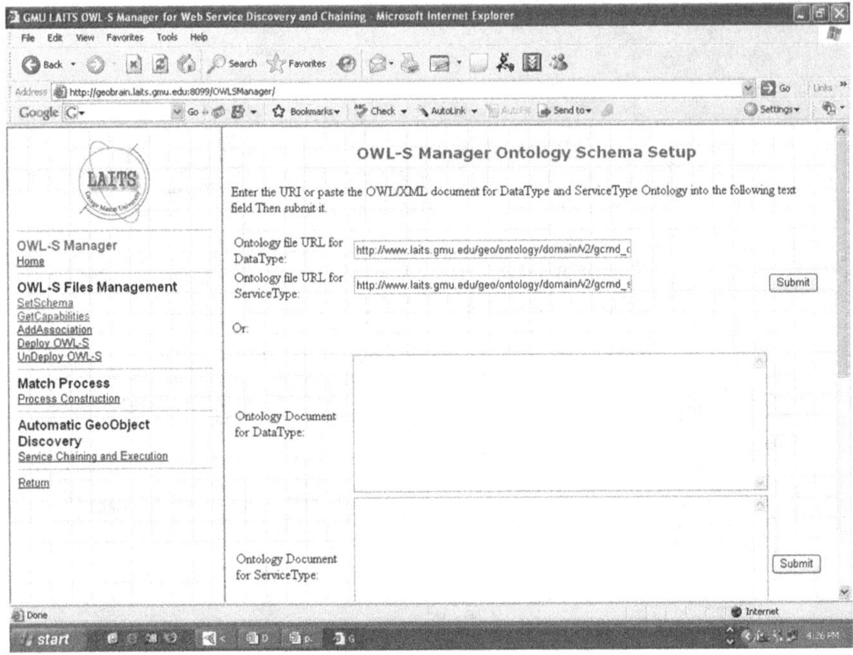

Fig. 8.4 Register geospatial DataType and ServiceType ontologies

b. OWL-S Deploy: Loads a service ontology (i.e. an OWL-S) to the knowledge base (Fig. 8.5).
c. OWL-S UnDeploy: Unloads a service ontology from the knowledge base (Fig. 8.6).
d. Get Capabilities: Gets the service ontology repository in the knowledge base (Fig. 8.7).

(2) The Semantic Matching function: Queries matched data and services;
(3) The Service Chaining function: Performs composition of geospatial services (Fig. 8.8).
(4) The chain execution function: Executes OWL-S composite processes resulting from the chaining process (Fig. 8.9).

In addition, for the path planning approach, additional JavaServer Pages (JSP) files are developed which can invoke the K-shortest path algorithm over the service graph generated from the knowledge base of OWLSManager (Fig. 8.10). A generation of Dijkstra's algorithm (Martins et al. 1998) is adopted for K-shortest path algorithm. Since it is a label setting algorithms (Martins et al. 1998), paths are determined throughout the computations instead of at the ending of algorithm, thus the efficiency is ensured when a large number of services are involved. The individual services with the same service profile can be selected based on the QoS

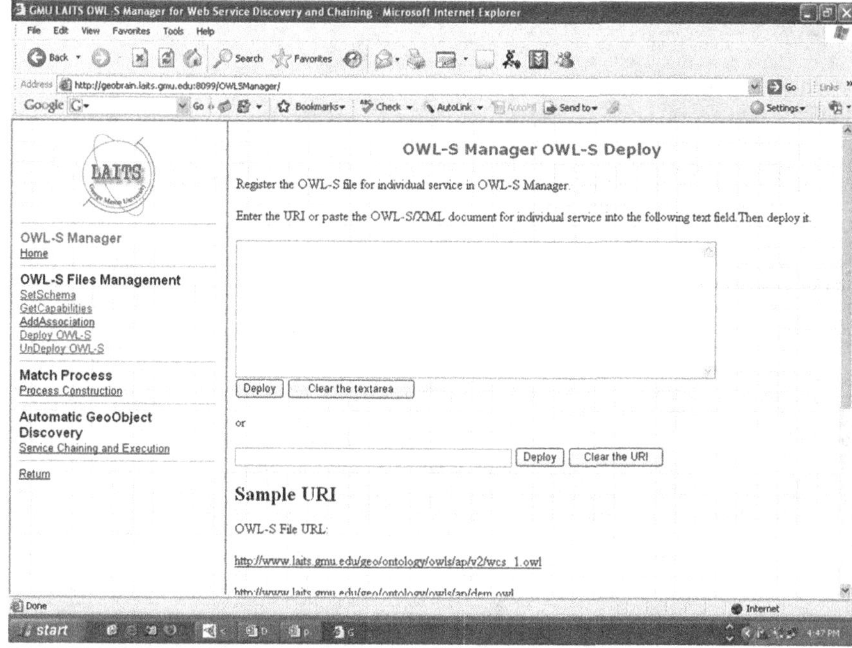

Fig. 8.5 Deploy OWL-S

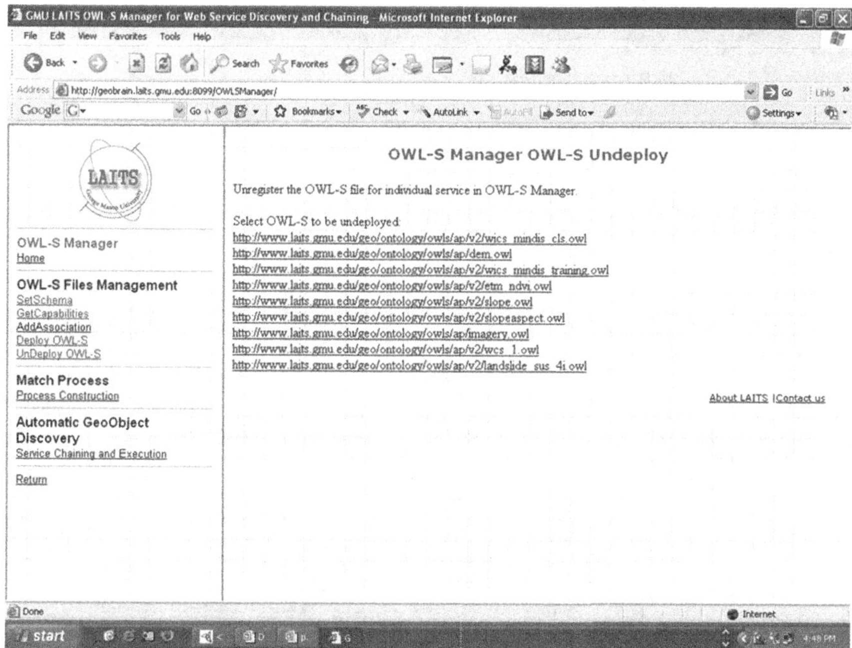

Fig. 8.6 Undeploy OWL-S

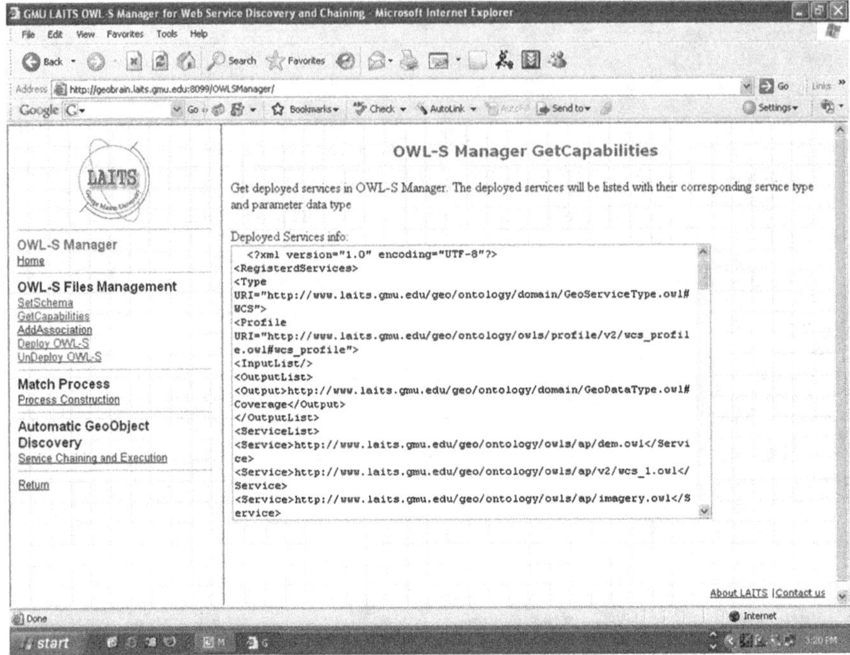

Fig. 8.7 The deployed OWL-Ss can be viewed using the GetCapabilities function

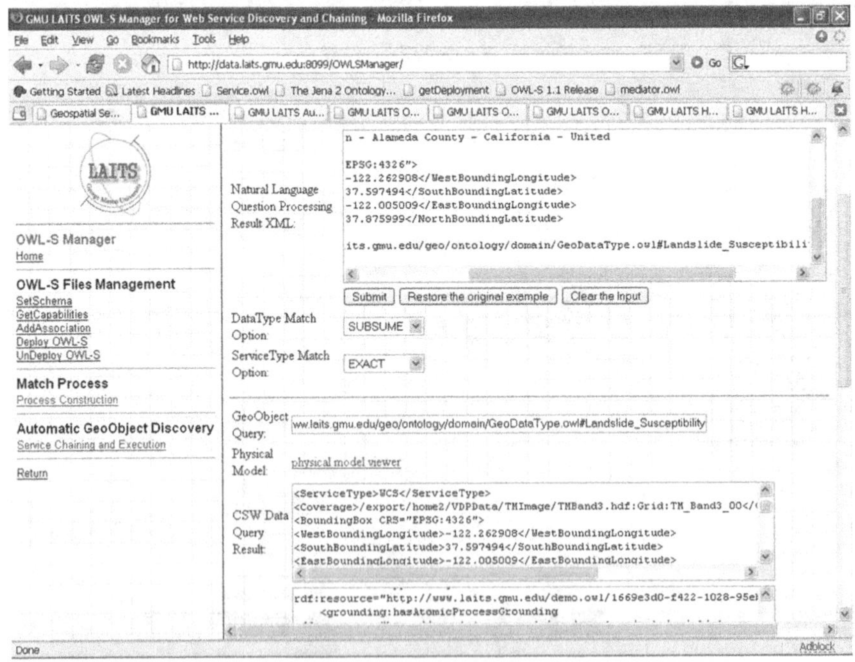

Fig. 8.8 Geospatial service composition

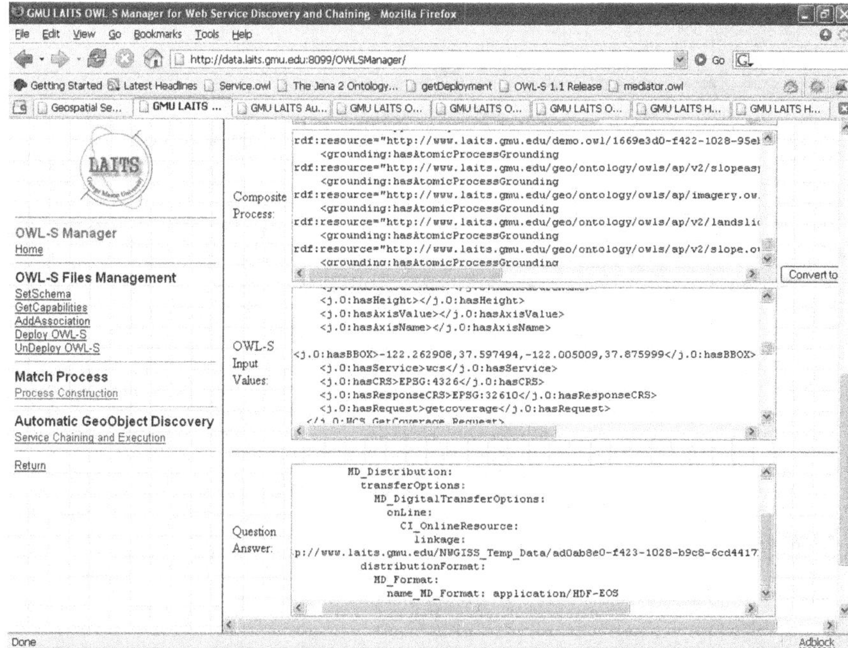

Fig. 8.9 Execution of composite process

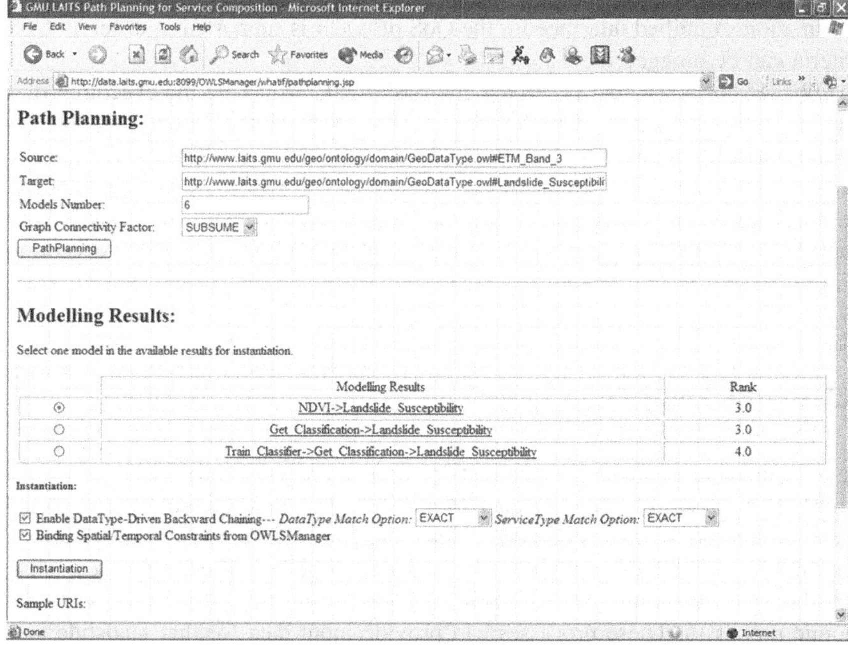

Fig. 8.10 The path planning user interface

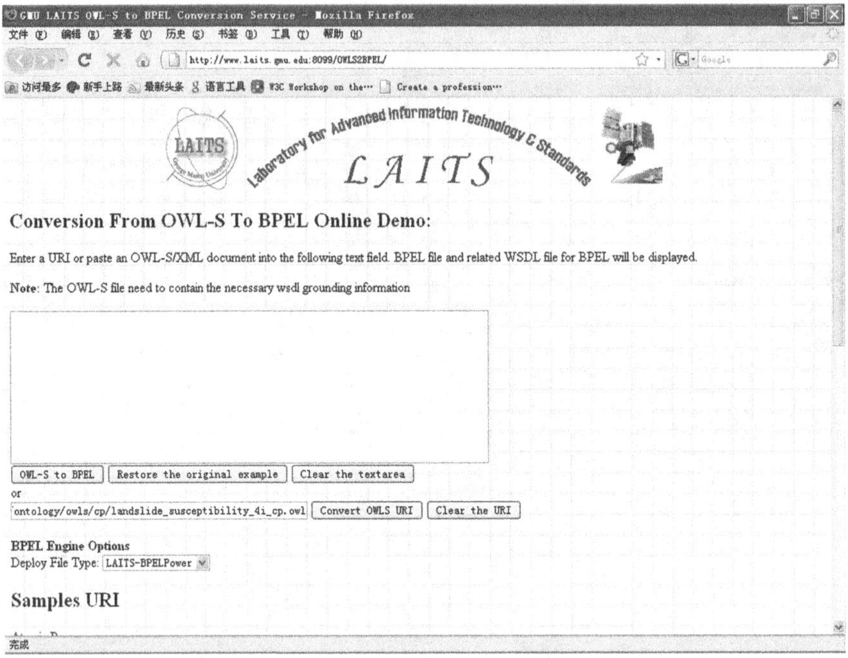

Fig. 8.11 OWL-S to BPEL conversion

information. A unified interface for the QoS provider is supported so different QoS criteria can be plugged in.

OWL-S's primary goal is service composition. Considering the execution of a service chain, a number of limitations in OWL-S have been identified, such as fault/error handling and event handling. These features are well defined in service composition languages such as BPEL. An OWL-S to BPEL conversion tool has been developed and implemented (Figs. 8.11, 8.12, and 8.13). It works as a Web application. The conversion results can be sent to a BPEL engine for execution (Fig. 8.14).

8.3 Result Analysis

In the landslide susceptibility case, two atomic process models can provide a landslide susceptibility data product (Fig. 8.15). The first atomic process model uses four types of data (slope, aspect, land cover, and NDVI) to calculate landslide susceptibility, and the second one only uses only slope and aspect data to calculate landslide susceptibility. Any of them can be combined with other discovered atomic processes. These processes can provide input data for that landslide susceptibility atomic process to create different composite process models. For

Fig. 8.12 Screen shot of OWL-S converted BPEL

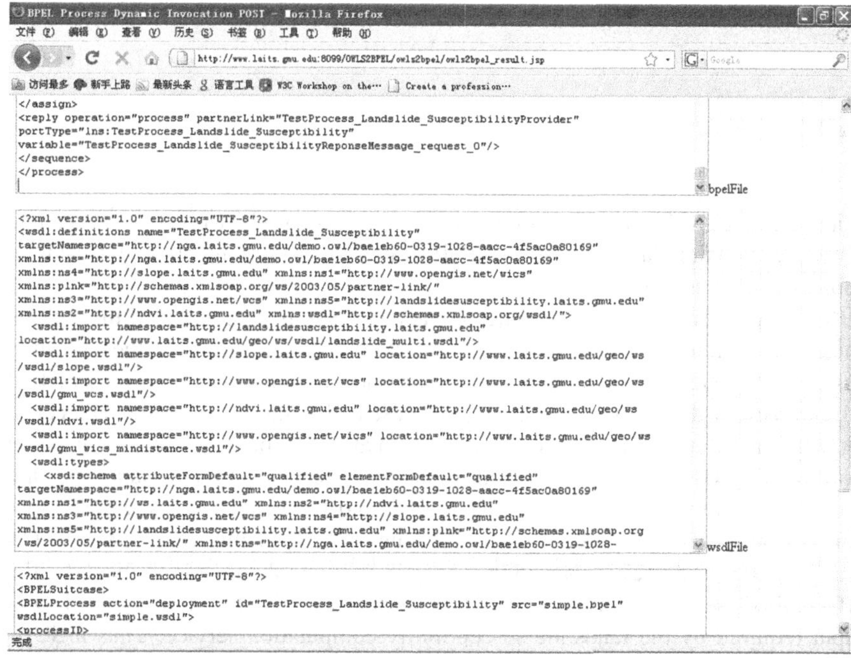

Fig. 8.13 Screen shot of OWL-S converted BPEL WSDL

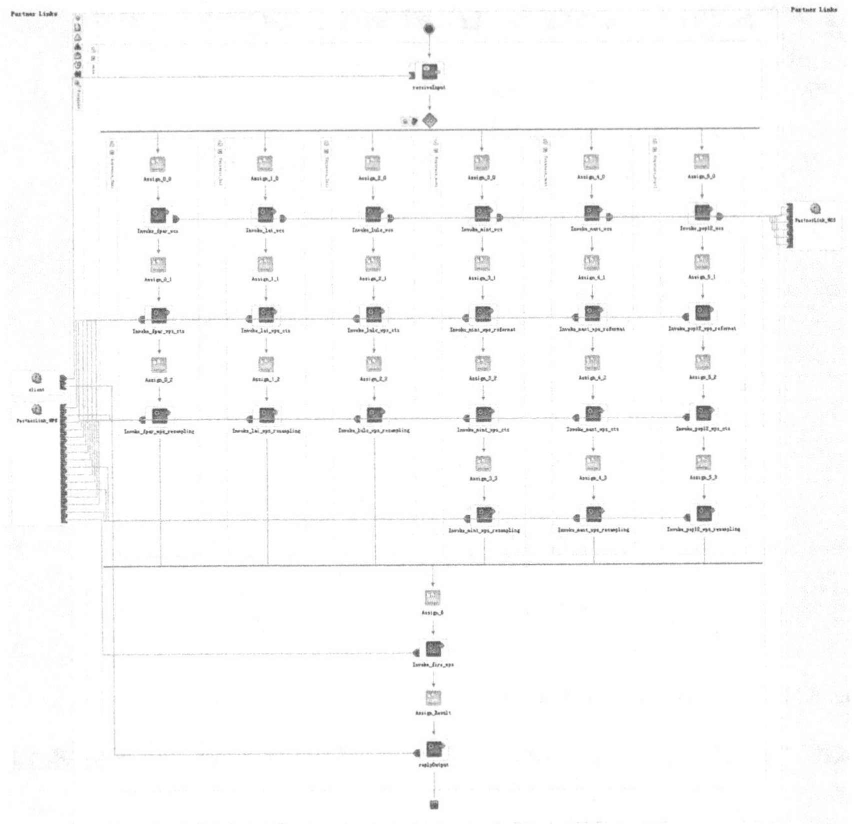

Fig. 8.14 A BPEL diagram

example, an ETM NDVI calculation process that provides ETM NDVI data can be linked to the first landslide susceptibility atomic process mentioned based on subsumption reasoning.

The applicability of the semantics-enhanced CSW is demonstrated through its support to automatic chaining of multiple Web services to derive the landslide susceptibility index of the certain area (Diamond Canyon, California) on a certain day. A virtual data product request is represented using a geospatial DataType along with spatial and temporal conditions. An atomic process model for landslide susceptibility (with either two or four inputs) is selected. Its input DataTypes are provided by slope and aspect (and land cover and NDVI) processes, whose input data are available and directly served through WCS. In this case, the EXACT match cannot produce the landslide susceptibility data automatically because the ETM NDVI service's output ETM NDVI is not exactly the same as the NDVI input required by the landslide susceptibility computation service. Thus a SUBSUME match is required to achieve this goal. If users select the RELAXED match

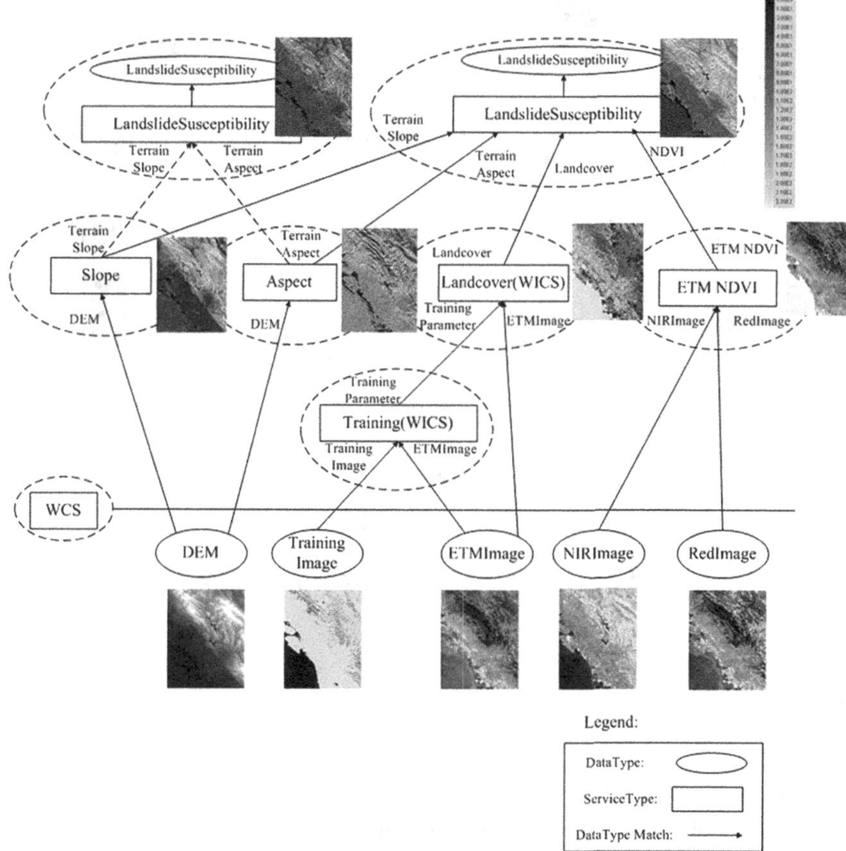

Fig. 8.15 Landslide susceptibility scenario

option, the goal still can be achieved since the match process considers the EXACT and SUBSUME matches first. The service chain in this use case can be automatically and dynamically generated whenever the CSW service is available and queries can be augmented with the semantic extensions in the ebRIM information model. The composite process can also be registered in the CSW as a virtual data product so that the composition process need not be repeated when a new request for the same data product is submitted. In addition to landslide susceptibility data, slope data, slope aspect data, landcover data or ETM NDVI data can also be created on demand. In this system, the service user is assisted by the ontologies from the knowledge base when selecting services and does not need to deal with syntactical service descriptions and WSDL message element mappings among possibly chainable services. Both OGC-compliant and non-OGC-compliant services are involved in the final service chain created by the automatic service composition process.

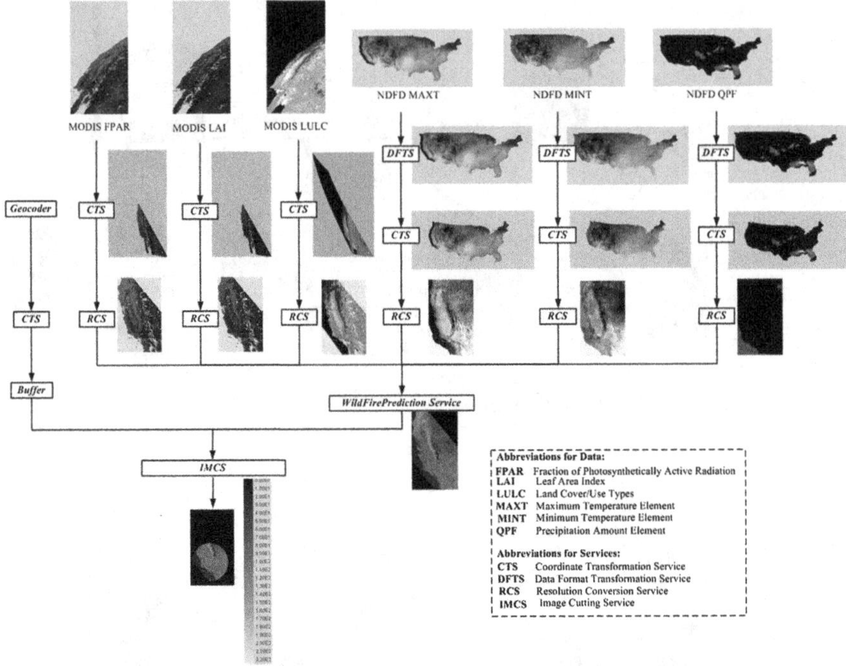

Fig. 8.16 Wildfire prediction scenario

To answer a "what if" question, for example, "what is the landslide risk for location L at time T if vegetation were changed?", the path planning method is applied to the use case of landslide risk again. The path modeling results from the source input "ETM Band 3" and the target output "Landslide Susceptibility" are shown in Fig. 8.10 to answer the example question. Each logical path is visualized as a linked graph created by WebDot.[1] The applicapability of this method are demonstrated by automatically chaining of multiple Web services to derive the landslide susceptibility index of the certain area (Diamond Canyon, California) on a certain day.

In the wildfire prediction scenario (Fig. 8.16), when formulating the goal, John needs to rely on the services to derive the bounding box for the desired wildfire data product, so that the query of the catalogue service can have the correct spatial filter. The client helps John to find the DistanceBuffer process. He appends a CTS process and a GetEnvelope process to get the geographic bounding box information. The role of the DistanceBuffer, CTS and GetEnvelope processes here is that of information-providing processes, which refers to services that can be used to gather information during the planning (Sirin et al. 2004).

[1] http://www.graphviz.org/.

In the first step of the wildfire case, the user relies on the information-providing processes to get the geographic bounding box as the spatial filter of CSW query. In the second step (process modeling), a lower-level model is generated by the plan generation component from the user's initial model. The user could also specify a wildfire prediction process with input and output geospatial "DataType"s in the initial model if the user is concerned with particular geospatial "DataType"s. In the third step (process instantiation), the user specifies the input geospatial "DataType"s of the lower-level process model that are needed to be bound (i.e. "Wildfire_Danger_Index"). The plan generation component interacts with the CSW and generates the physical model. In the physical model, NOAA NDFD data and NASA MODIS data are bound to the service chain. In the wildfire prediction service, eighteen preconditions are defined. They cover the six input "DataType"s and three types of different metadata entities: the file format, coordinate spatial reference system, and grid resolution. Failure to satisfy any precondition leads to the insertion of one data reduction and transformation service, following the domain procedures of the metadata-tracking component. Thus, the physical model is transformed to an executable workflow. The result is sent to the chain execution component for execution. In all these steps, the user is assisted by the ontologies from the knowledge base. Therefore, the user has been released from the heavy syntactical service descriptions and message element mappings between the chainable services.

The system can help the domain expert focus more on the domain knowledge contribution instead of delving into the technical details. The prototype system demonstrates that both individual geoprocessing services and valid process models can be shared. The system is thus a self-evolving system whose capability will increase significantly as more individual services and modeling processes are inserted and/or developed, thus it contributes to the evolution of the Cyberinfrastructure.

Therefore, the use cases help to demonstrate that (a) interoperable geospatial Web services can be used to discover, request, access, and obtain geospatial data and information in a distributed environment; (b) such services can be intelligently orchestrated to generate geospatial process model that transform data into information and information into knowledge which assists decision makings; (c) the geospatial process models can be automatically converted to executable service chains and be invoked and executed on demand; (d) the process models and service chains can be archived, catalogued, and be advertised as new geospatial services and thus be discovered and used in future geospatial modeling; and (e) such intelligent Web service based modeling processes will maximize the potential of individual data and services and significantly advance the geospatial knowledge discovery.

It is noted that the performance is not the main concern of the work. This work focuses on the representation of semantics for geospatial data and Web service, exploring its usage in the geospatial service composition, and the design of a general framework. It takes advantage of the third-party developed tool for the most time-consuming work during the generation of executable workflow, namely, using Jena as the reasoner and query engine. For example, the running time of two typical examples are compared. The first example is used to demonstrate the subsumption reasoning to get the semantically matched Web services in the service composition process. Landslide susceptibility data product is produced through the service composition upon the request. Furthermore, the second example demonstrates the mediated RDF structure as the metadata relay structure strategy in the service composition process to facilitate the metadata tracking and built-in rule usage in the OWL-S preconditions satisfiability. Two preconditions including the data format and coordinate reference system are defined in the OWL-S for slope service. The experiments ran on a Linux -server with 2600 MHz Intel Pentium processor, 512 MB of RAM, and under a Ubuntu operating system with tomcat Web server. The average running time of the first example is 27 s while the second example is 65 s. Compared with the first example 1, the second example takes more time. It is mainly due to two reasons: (1) although the first involves much more services, it is a parallel computation. (2) OWL-S files in the second example are assigned with OWL-S preconditions. It shows that precondition checking and SWRL processing consumes additional time, especially when ABOX reasoning is involved in the knowledge base (11178 statements). This implies that the performance of the current implementation highly depends on the performance of the reasoning engine.

Chapter 9
Conclusions

This book addresses the key technologies for developing an intelligent geospatial knowledge system based on Web services. In particular, Semantic Web technologies are introduced into the area of geospatial Web services. The ontology-based approach helps to improve the recall and precision of data and services discovery provided by the catalogue service. Semantics-enabled metadata tracking and satisfaction allows analysts to focus on the generation of a geospatial process model rather than spending large amounts of time in data preparation. "DataType"-driven service composition and path planning can help to automate a range of knowledge discovery processes in a limited geospatial domain. Process planning facilitates the construction of complex services and models for geocomputation. The process models and service chains can be archived, catalogued, and advertised as new geospatial services in CSW and thus be discovered and used for future geospatial modeling.

Semantics for geospatial data, services, and service chains, are organized and registered in the CSW by extending ebRIM elements. In particular, the IOPE semantics are introduced for geospatial services and the registration of IOPE semantics is proposed in the CSW-ebRIM profile for loosely-coupled, mixed-coupled, and tightly-coupled services and data. A ProcessModel class is added to the ebRIM model. It is a key extension, addressing the requirements for geospatial information processing and knowledge discovery purpose in a data-rich distributed environment, as opposed to the initially core extension class Dataset in the CSW-ebRIM profile. Middleware to support semantics-enhanced discovery of geospatial data, services/service chains, and process models has been developed. Such middleware can be applied to support materialization of virtual data products. The approach demonstrates that semantics can be used to improve the data/services/service chains discovery capability of geospatial catalogue services. The process models, working as a kind of geospatial knowledge, can address the analysis issues in the Cyberinfrastructure and support on-demand delivery of geospatial information and knowledge.

The book shows that standards-based geospatial Web services can be used to discover, request, access, and obtain geospatial data and information in a

P. Yue, *Semantic Web-based Intelligent Geospatial Web Services*,

SpringerBriefs in Computer Science, DOI: 10.1007/978-1-4614-6809-7_9,

distributed environment. A general procedure for automatic geospatial service composition is proposed. This procedure includes three phases: process modeling, process model instantiation, and workflow execution. The approach in this book can address all types of service chaining in the OGC abstract service architecture: user-defined, workflow-managed and aggregate. The existing process models allow analysts to interactively construct new, complex Web-executable geospatial process models.

Future work includes the application of the approach in various use cases involving more types of geoprocessing services and generation of more complex metadata. Such an application should provide primarily semantic descriptions of all involved services. For example, over one hundred geoprocessing services have been developed in GeoPW (Yue et al. 2010). How these services determine the appropriateness of starting data or affect the input data should be described semantically, explicitly, and formally. And we believe that specifying such metadata requirements and metadata changes by an appropriate semantic description of geoprocessing services is important to allowing automatic service chaining and semantic evaluation of service chains in the future. In addition, the insertion of data reduction and transformation services for satisfying metadata constraints is implemented as domain control procedures by providing functions in the computer source code. These domain control procedures embody domain knowledge, which can be expressed using some knowledge representation mechanism to facilitate AI planning. It is possible to transform the problem of satisfying metadata constraints to an AI planning domain so that traditional AI planners can be used with semantic geospatial services to generate a plan for service chains.

The CSW-ebRIM profile has been adopted because it allows catalogues to handle geospatial data, services, and other types of information resources such as applications schemas, software components, and reference documents; also, it can handle process models, as shown in this chapter. This capability to register different information resources is demonstrated through the employment of those extensibility points in this paper. There are other catalogue services, for example, the NASA EOS Clearinghouse (ECHO), the U.S. Department of Energy (DOE) Earth System Grid (ESG) Simulation Data Catalogue, discovery frameworks like UDDI, and other application profiles of the OGC catalogue specification like the ISO metadata application profile. Some of them differ in the catalogue query language and communication protocol, while others differ in the information model. If multiple catalogue services must be used, they can be federated to provide a comprehensive discovery of geospatial information. It is possible to add the semantic middleware in the federation service as a future work.

Bibliography

Aalst W (2003) Don't go with the flow: web services composition standards exposed. IEEE Intell Syst 2003:72–76

Aalst W, Hofstede A (2004) YAWL: yet another workflow language. Inf Syst 30(4):245–275

Aissi S, Malu P, Srinivasan K (2002) E-business process modeling: the next big step. IEEE Comput 35(5):55–62

Akkiraju R et al (2005) Web service semantics—WSDL-S http://www.w3.org/2005/04/FSWS/Submissions/17/WSDL-S.html

Akkiraju R, Goodwin R, Doshi P, Roeder S (2003) A method for semantically enhancing the service discovery capabilities of UDDI. In: Proceedings of the workshop on information integration on the web, eighteenth international joint conference on artificial intelligence (IJCAI), Mexico, pp 87–92

Akram A, Meredith D, Allan R (2006) Evaluation of BPEL to scientific workflows. In: Proceedings 6th IEEE international symposium on cluster computing and the grid, pp 269–274

Alameh N (2003) Chaining geographic information web services. IEEE Internet Comput 07(5):22–29

Altintas I, Birnbaum A, Baldridge K, Sudholt W, Miller M, Amoreira C, Potier Y, Ludäscher B (2004) A framework for the design and reuse of grid workflows. International workshop on scientific applications on grid computing, SAG 2004, LNCS 3458, Springer, pp 119–132

Antoniou G, Harmelen Fv (2004) A semantic web primer. The MIT Press, Cambridge, pp 17–18

Baader F, Nutt W (2003) Basic description logics. In: Baader F, Calvanese D, McGuinness D, Nardi D, Patel-Schneider P (eds) The description logic handbook. Theory, implementation and applications. Cambridge University Press, Cambridge, pp 43–95

Battle R, Kolas D (2012) Enabling the geospatial semantic web with Parliament and GeoSPARQL. Seman Web J. doi:10.3233/SW-2012-0065

Baumann P (2010) OGC WCS 2.0 Interface Standard—Core, Version 2.0.0, OGC 09-110r3, Open Geospatial Consortium, Inc., p 53

Bechini A, Tomasi A, Viotto J (2008). Enabling ontology-based document classification and management in ebXML registries, In: Proceedings of the 2008 ACM symposium on applied computing, Fortaleza, Ceara, pp 1145–1150

Benatallah B, Dumas M, Fauvet M-C, Abhi FA (2001) Towards patterns of web service composition. Technical report, UNSWCSE -TR-0111, University of New South Wales, p 35

Bermudez EL (2004) Ontomet: ontology metadata framework. Ph.D. Dissertation, Drexel University, Philadelphia, p 177

Berners-Lee T, Hendler J, Lassila O (2001) The semantic web. Sci Am 284(5):34–43

Berners-Lee T (2000a) Semantic web talk. Invited Talk at XML 2000 Conference, Slides: http://www.w3.org/2000/Talks/1206-xml2k-tbl/slide10-0.html

P. Yue, *Semantic Web-based Intelligent Geospatial Web Services*, SpringerBriefs in Computer Science, DOI: 10.1007/978-1-4614-6809-7,

Berners-Lee T (2000b) CWM—closed world machine. Internet: http://www.w3.org/2000/10/swap/doc/cwm.html

Berners-Lee T (1998) Semantic Web road map. Internet: http://www.w3.org/DesignIssues/Semantic.html

Berrick S, Leptoukh G, Farley J, Rui H (2009) Giovanni: a web service workflow-based data visualization and analysis system. IEEE Trans Geosci Remote Sens 47(1):106–113

Bizer C, Heath T, Idehen K, Berners-Lee T (2008) Linked data on the web. In: Proceedings of linked data on the web workshops, WWW 2008, Beijing, pp 1265–1266

Boley H, Tabet S, Wagner G (2001) Design rationale of RuleML: a markup language for semantic web rules, Proceedings of SWWS'01. Stanford, July/Aug 2001. p 21

Booth D, Haas H, McCabe F, Newcomer E, Champion M, Ferris C, Orchard D (2004). Web services architecture. W3C Working Group Note 11 Feb 2004, W3C, http://www.w3.org/TR/ws-arch/

Bowers S, Ludäscher B (2004) An ontology-driven framework for data transformation in scientific workflows. In: Proceedings of the international workshop on data integration in the life sciences (DILS), vol 2994 of LNCS, Springer, pp 1–16

Brachman RJ, Levesque HJ (2003) Knowledge representation and reasoning. Elsevier, pp 314–316

Brodaric B, Fox P, McGuinness DL (2009) Editorial: geoscience knowledge representation in cyberinfrastructure. Comput Geosci 35(4):697–868

Bröring A, Stasch C, Echterhoff J (2012) OpengGIS® sensor observation service interface standard, Version 2.0, OGC 12-006, Open Geospatial Consortium, Inc., p 163

Bruijn J de et al (2005) Web service modeling ontology (WSMO). http://www.w3.org/Submission/WSMO/

Cardoso J, Sheth A (2003) Semantic e-workflow composition. J Intell Inf Syst 21(3):191–225

Cardoso J, Sheth A (2005) Introduction to semantic web services and web process composition. In: Cardoso J, Sheth A (eds) First international workshop on semantic web services and web process composition (SWSWPC 2004), LNCS 3387. Springer, Berlin, p 14

Casati F, Ilnicki S, Jin L (2000) Adaptive and dynamic service composition in EFlow. In: Proceedings of 12th international conference on advanced information systems engineering(CAiSE), Stockholm, June 2000, Springer

Casati F, Sayal M, Shan M (2001) Developing e-services for composing e-services. In: Proceedings of 13th international conference on advanced information systems engineering(CAiSE), Interlaken, Switzerland, June 2001. Springer, p 27

Christensen E, Curbera F, Meredith G, Weerawarana S (2001) Web services description language (WSDL) 1.1. http://www.w3.org/TR/2001/NOTE-wsdl-20010315

Clark J (1999) XSL transformations (XSLT). W3C recommendation, 16 Nov 1999, W3C, http://www.w3.org/TR/xslt

Clark J, DeRose S (1999) XML path language. W3C recommendation, 16 Nov 1999, W3C, http://www.w3.org/TR/xpath

Clery D, Voss D (2005) All for one and one for all. Science 308(5723):809

Colgrave J, Akkiraju R, Goodwin R (2004) External matching in UDDI. IEEE international conference on web services, San Diego, 2004, p 8

Dean M, Schreiber G (eds) (2004). OWL web ontology language reference, W3C. http://www.w3.org/TR/owl-ref

de la Beaujardière J (2006) OpenGIS web map server implementation specification. Version 1.3.0, OGC 06-042, Open Geospatial Consortium, Inc., p 85

Di L (2005a) A framework for developing web-service-based intelligent geospatial knowledge systems. J Geogr Inf Sci 11(1):24–28

Di L, Zhao P, Yang W, Yu G, Yue P (2005b) Intelligent geospatial web services. IEEE international geoscience and remote sensing symposium, 2005. IGARSS '05. Proceedings, vol 2, July 2005, pp 1229–1232

Di L (2004) GeoBrain-a web services based geospatial knowledge building system. Proceedings of NASA earth science technology conference 2004. June 22–24, 2004. Palo Alto, p 8(CD-ROOM)

Di L, McDonald K (1999) Next generation data and information systems for earth sciences research. In: Proceedings of the first international symposium on digital earth, vol I. Science Press, Beijing, pp 92–101

Dogac A, Kabak Y, Laleci GB (2004) Enriching ebXML registries with OWL ontologies for efficient service discovery. In: Proceedings of the 14th international workshop on research issues on data engineering: web services for E-commerce and E-government applications (RIDE' 04), Boston, pp 69–76

Dogac A, Kabak Y, Laleci GB, Mattocks C, Najmi F, Pollock J (2005) Enhancing ebXML registries to make them OWL aware. Distrib Parallel Databases J 18(1):9–36

Dogac A (ed) (2006). ebXML registry profile for web ontology language (OWL), Version 1.5, regrep-owl-profile-v1.5-cd01, Organization for the Advancement of Structured Information Standards (OASIS), p 76

Drexel (2004) ISO 19115 metadata ontology. Drexel University. http://loki.cae.drexel.edu/~wbs/ontology/. Accessed 17 Oct 2005

EarthCube, EarthCube Capabilities, Dec 2011. http://earthcube.ning.com/page/capabilities

Egenhofer M (2002) Toward the semantic geospatial web. In: The 10th ACM international symposium on advances in geographic information systems (ACM-GIS), McLean, p 4

ESA (2012) European Space Agency (ESA) grid processing on demand (G-POD) for earth observation applications. http://gpod.eo.esa.int/. Accessed 24 March 2012

Farrell J, Lausen H (2006) Semantic annotations for WSDL (SAWSDL). http://www.w3.org/TR/sawsdl/

Fensel D, Bussler C (2002) The web service modeling framework WSMF. Technical report, Vrije Universiteit Amsterdam, 2002, p 33

Fonseca F, Sheth A (2002) The geospatial semantic web. University consortium for geospatial information science (UCGIS) research priorities. USA, p 2

Foster I (2005) Service-oriented science. Science 308(5723):814–817

Friis-Christensen A, Lucchi R, Lutz M, Ostlaender N (2009) Service chaining architectures for applications implementing distributed geographic information processing. Int J Geogr Inf Sci 23(5):561–580

GENESI-DEC (2012) The European Commission Ground European Network for Earth Science Interoperations-Digital Earth Communities (GENESI-DEC project). http://www.genesi-dec.eu. Accessed 24 March 2012

GEON (2003) A research project to create cyber infrastructure for the geosciences. NSF/ITR. www.geongrid.org. Accessed 24 March 2012

GEOSS (2012) Geo portal. http://www.geoportal.org/web/guest/geo_home. Accessed 24 March 2012

Gomadam K, Ranabahu A, Sheth A (2010) SA-REST: semantic annotation of web resources, W3C Member Submission. http://www.w3.org/Submission/SA-REST/. Accessed April 2010

Gruber TR (1993) A translation approach to portable ontology specification. Knowl Acquis 5(2):199–220

Hatzi O, Vrakas D, Nikolaidou M, Bassiliades N, Anagnostopoulos D, Vlahavas I (2012) An integrated approach to automated semantic web service composition through planning. IEEE Trans Serv Comput 5(3):319–332

Hey T, Trefethen AE (2005) Cyberinfrastructure for e-Science. Science 308(5723):817–821

Horrocks I, Patel-Schneider PF, Boley H, Tabet S, Grosof B, Dean M (2004) SWRL: A semantic web rule language combining OWL and RuleML. W3C Member Submission. (2004) http://www.w3.org/Submission/SWRL/

HP (2006) Jena. Hewlett-Packard Labs Semantic Web Programme. http://jena.sourceforge.net. Accessed 19 Nov 2009

ISO/TC211 (2007) Geographic information/geomatics. http://www.isotc211.org/

ISO/TC211 (2003) ISO19115:2003, Geographic information—metadata

ISO/TC211 (2005) ISO19119:2005, Geographic information—services

Jaeger E, Altintas I, Zhang J, Ludäscher B, Pennington D, Michener W (2005) A scientific workflow approach to distributed geospatial data processing using web services. In: 17th international conference on scientific and statistical database management (SSDBM'05), 27–29 June 2005, Santa Barbara, California. pp 87–90

Kammersell W, Dean M (2006) Conceptual search: incorporating geospatial data into semantic queries. In: Terra Cognita 2006, Workshop of 5th international semantic web conference. 5–9 Nov 2006, Athens, Georgia, p 10

Kifer M (2008) Rule interchange format: the framework. In: Web reasoning and rule systems. Lecture Notes in Computer Science, vol 5341, Springer, Berlin, pp 1–11

Kim J, Gil Y, Ratnakar V (2006) Semantic metadata generation for large scientific workflows. 5th international semantic web conference. 5–9 Nov 2006, Athens, LNCS 4273

Klusch M, Gerber A, Schmidt M (2005) Semantic web service composition planning with OWLS-Xplan, agents and the semantic web, 2005 AAAI Fall Symposium Series, Arlington Virginia, Nov 2005, p 8

Klyne G, Carroll JJ (2004) Resource description framework (RDF): concepts and abstract syntax. W3C Recommendation. http://www.w3.org/TR/2004/REC-rdf-concepts-20040210/. Accessed 10 Feb 2004

Kolas D, Hebeler J, Dean M (2005) Geospatial semantic web: architecture of ontologies. In: Proceedings of first international conference on geospatial semantics (GeoS 2005). Mexico City, Springer, pp 183–194

Kolas D, Dean M, Hebeler J (2006) Geospatial Semantic Web: architecture of ontologies. In: Proceedings of 2006 IEEE aerospace conference. 4–11 March 2006, Big Sky, Montana, p 10

Lemmens R, Wytzisk A, By Rd, Granell C, Gould M, van Oosterom P (2006) Integrating semantic and syntactic descriptions to chain geographic services. IEEE Internet Comput 10(5):18–28

Lieberman J, Pehle T, Dean M (2005) Semantic evolution of geospatial web services: use cases and experiments in the geospatial semantic web. Talk at the W3C Workshop on Frameworks for Semantic in Web Services, Innsbruck, Austria. p 20

Liu W, He K, Liu W (2005) Design and realization of ebXML registry classification model based on ontology. In: Proceedings of the international conference on information technology: coding and computing (ITCC'05), pp 809–814

Lucchi R, Millot M, Elfers C (2008) Resource oriented architecture and REST. Technical report, European Commission, Joint Research Centre, p 16

Ludäscher B, Altintas I, Berkley C, Higgins D, Jaeger E, Jones M, Lee E, Tao J, Zhao Y (2005) Scientific workflow management and the kepler system. Concurrency and computation: practice & experience, Special Issue on Scientific Workflows, to appear, 2005, p 19

Lutz M, Kolas D (2007) Rule-based discovery in spatial data infrastructures. Transactions in GIS, Special Issue on the Geospatial Semantic Web, in press. p 27

Lutz M (2004) Non-taxonomic relations in semantic service discovery and composition. In: Maurer F, Ruhe G (eds) Proceedings of the first "ontology in action" workshop, in conjunction with the sixteenth international conference on software engineering & knowledge engineering (SEKE'2004). pp 482–485

Lutz M, Klien E (2006) Ontology-based retrieval of geographic information. Int J Geogr Inf Sci 20(3):233–260

Mabrouk M (ed) (2003) OpenGIS@ location services (OpenLS): core services. Version: 1.0. OGC 03-006r3. Open Geospatial Consortium, Inc., p 165

Martell R (2008) CSW-ebRIM Registry Service—Part 1: ebRIM profile of CSW, Version 1.0.0, OGC 07-110r2, Open Geospatial Consortium, Inc., p 57

Martin D et al (2004) OWL-based web service ontology (OWL-S). http://www.daml.org/services/owl-s/1.1

Martins EQV, Pascoal MMB, Santos JLE (1998) The K shortest paths problem. Research Report, CISUC, June 1998, p 21

Maue P, Michels H, Roth M (2012) Injecting semantic annotations into (geospatial) web service descriptions. Seman Web J 3(4)

Medjahed B, Bouguettaya A, Elmagarmid AK (2003) Composing web services on the semantic web. VLDB J 12(4):333–351

MINDSWAP (2004) OWL-S API. Maryland Information and Network Dynamics Lab Semantic Web Agents Project (MINDSWAP). http://www.mindswap.org/2004/owl-s/api/. Accessed 19 Nov 2009

Müller MU, Warmerdam F, Whiteside A, Fellah S (2004) OpenGIS® web coordinate transformation service (WCTS) implementation specification, Version: 0.1.9, OGC 04-0XX, Open Geospatial Consortium, Inc., p 116

Nebert D, Whiteside A, Vretanos P (eds) (2007) OpenGIS@ catalog services specification. Version 2.0.2, OGC 07-006r1, Open GIS Consortium Inc., p 218

NSF (2007) Cyberinfrastructure vision for 21st century discovery, National Science Foundation, USA, March 2007, p 64

OASIS (2004) The UDDI technical white paper. http://uddi.org/pubs/uddi-tech-wp.pdf

OASIS (2005) ebXML Registry Information Model Version 3.0, OASIS Standard, 2 May, 2005. regrep-rim-3.0-os, p 78

OASIS (2007) Web services business process execution language, version 2.0. Web services business process execution language (WSBPEL) Technical Committee (TC), p 264

Olsen LM, Major G, Leicester S, Shein K, Scialdone J, Weir H, Ritz S, Solomon C, Holland M, Bilodeau R, Northcutt T, Vogel R (2004) NASA/Global Change Master Directory (GCMD) Earth Science Keywords. Version 5.1.1. http://gcmd.nasa.gov/Resources/valids/keyword_list.html

Paolucci M, Kawamura T, Payne TR, Sycara K (2002a) Importing the semantic web in UDDI. In Web SERVICES, E-Business and Semantic Web Workshop, 2002

Paolucci M, Kawamura T, Payne TR, Sycara K (2002b) Semantic matching of web services capabilities. In: Horrocks I, Hendler JA (eds) The semantic web-ISWC 2002, First international semantic web conference, Sardinia, June 9–12, 2002, Proceedings. Lecture Notes in Computer Science 2342, Springer, Berlin, 2002, pp 333–347

Pautasso C, Zimmermann O, Leymann F (2008) RESTful web services vs. big web services: Making the right architectural decision. In: Proceedings of the 17th world wide web conference, Beijing, China, 2008, pp 805–814

Papazoglou MP (2003) Service-oriented computing: concepts, characteristics and directions. Keynote for the 4th international conference on web information systems engineering (WISE 2003). pp 3–12

Peer J (2005) Web service composition as AI planning—a survey. University of St.Gallen, Switzerland, p 63

Percivall G (ed) (2002) The OpenGIS abstract specification, topic 12: OpenGIS Service Architecture, Version 4.3, OGC 02-112. Open Geospatial Consortium, Inc., p 78

Ponnekanti SR, Fox A (2002) SWORD: a developer toolkit for web service composition. In: Proceedings of the international world wide web conference, Honolulu, Hawaii, May 2002, pp 83–107

Portele C (ed) (2007) OpenGIS geography markup language (GML) encoding standard, Version 3.2.1. OGC 07-036, Open Geospatial Consortium, Inc., p 437

Prud'hommeaux E, Seaborne A (2006) SPARQL query language for RDF. W3C Working Draft. http://www.w3.org/TR/rdf-sparql-query/. Accessed 4 Oct 2006

Rao J, Su X (2004) A survey of automated web service composition methods. Proceedings of the first international workshop on semantic web services and web process composition, SWSWPC 2004, California, pp 43–54

Raskin R, Pan M (2005) Knowledge representation in the semantic web for earth and environmental terminology (SWEET). Comput Geosci 31(9):1119–1125

Rodriguez MA, Egenhofer MJ (2003) Determining semantic similarity among entity classes from different ontologies. IEEE Trans Knowl Data Eng 15(2):442–456

Roman D, Klien E, Skogan D (2006) SWING—a semantic web services framework for the geospatial domain, Terra Cognita 2006. International Semantic Web Conference ISWC'06 Workshop, 5–9 Nov 2006. Athens, Georgia

Russel S, Norvig P (2003) Artificial intelligence: a modern approach, 2nd edn. Prentice-Hall Inc.

Sarkar S, Kanungo DP (2004) An integrated approach for landslide susceptibility mapping using remote sensing and GIS. Photogramm Eng Remote Sens 70(5):617–625

Schut P (2007) OpenGIS® web processing service, version 1.0.0, OGC 05-007r7, Open Geospatial Consortium, Inc., p 87

Sheth A (2003) Semantic web process lifecycle: role of semantics in annotation, discovery, composition and orchestration, invited talk at WWW 2003 Workshop on E-Services and the Semantic Web. Budapest, Hungary, p 63

Sheth A (1999) Changing focus on interoperability in information systems: from system, syntax, structure to semantics. In: Goodchild MF, Egenhofer M, Fegeas R, Kottman CA (eds) Interoperating geographic information systems. Kluwer, New York, pp 5–30

Sirin E, Parsia B, Wu D, Hendler J, Nau D (2004) HTN planning for web service composition using SHOP2. J Web Seman 1(4):377–396

Sirin E, Hendler J, Parsia B (2003) Semi-automatic composition of web services using semantic descriptions. In: Proceedings of workshop on web services: modeling, architecture and infrastructure (WSMAI), ICEIS Press, pp 17–24

Sivashanmugam K, Verma K, Sheth AP, Miller JA (2003) Adding semantics to web services standards. In: 1st proceedings of the international conference on web services, ICWS '03. Las Vegas, Nevada, p 7

Sonnet J (ed) (2005) OWS 2 common architecture: WSDL SOAP UDDI. Version: 1.0.0. OGC 04-060r1. Open Geospatial Consortium, Inc., p 76

Srinivasan N, Paolucci M, Sycara K (2004) Adding OWL-S to UDDI, implementation and throughput. First international workshop on semantic web services and web process composition, San Diego, p 12

Stollberg B, Zipf A (2007) OGC web processing service interface for web service orchestration—aggregating geo-processing services in a bomb threat scenario. In: Proceedings 7th international symposium of web and wireless geographical information systems (W2GIS 2007), Cardiff, Lecture Notes in Computer Science (LNCS) 4857, pp 239–251

SWSI (2004) Semantic Web Services Initiative (SWSI). http://www.swsi.org/

Tatem AJ, Goetz SJ, Hay SI (2008) Fifty years of earth observation satellites. Am Sci 96(5):390–398

Vretanos P (2010) OpenGIS Web Feature Service 2.0 Interface Standard. Version 2.0.0, OGC 09-025r1, Open Geospatial Consortium, Inc., p 253

W3C (2001) Semantic Web. http://www.w3.org/2001/sw/. Accessed Feb 2007

W3C (2007a) Simple object access protocol(SOAP) specifications. World Wide Web Consortium (W3C). http://www.w3.org/TR/soap/. Accessed 18 Oct 2012

W3C (2007b) Web services description language (WSDL) 2.0, World Wide Web Consortium (W3C). http://www.w3.org/TR/2007/REC-wsdl20-adjuncts-20070626/#_http_binding_default_rule_method. Accessed 16 Oct 2011

Wei Y, Di L, Zhao B, Liao G, Chen A, Bai Y, Liu Y (2005) The design and implementation of a grid-enabled catalogue service. In: 25th anniversary of IEEE international geoscience and remote sensing symposium (IGARSS 2005), 25–29 July, COEX, Seoul, pp 4224–4227

WfMC (1999) Workflow management coalition, terminology & glossary. Document Number WFMC-TC-1011. http://www.wfmc.org/standards/docs/TC-1011_term_glossary_v3.pdf, p 65

WfMC (2008) Process definition interface—XML process definition language. Workflow Management Coalition, Cohasset, p 217

Yang W, Whiteside (2005) Web image classification service (WICS) implementation specification. Version: 0.3.3. OGC 05-017. Open Geospatial Consortium, Inc., p 45

Yue P, Gong J, Di L, Yuan J, Sun L, Sun Z, Wang Q (2010) GeoPW: laying blocks for the geospatial processing web. Trans GIS 14(6):755–772

Zaharia R, Vasiliu L, Hoffman J, Klien E (2009) Semantic execution meets geospatial web services: a pilot application. Trans GIS 12(1):59–73